ADVANCE PRAISE FOR

Media and Transnational Climate Justice

"*Media and Transnational Climate Justice* puts indigenous voices at the center of how we understand climate justice, offering an expansive analysis of transnational solidarity and the ways that it shapes and is shaped by sophisticated forms and practices of media activism. By combining the experiences of indigenous media activists, analysis of indigenous representations in legacy news, and fresh theoretical insight that challenge dominant climate politics, this exceptionally thoughtful and well researched book offers a blueprint for media justice. It is an essential read not only for those seeking to understand and reform climate politics but also for those interested in the ways media can support rather than undermine justice."—*Adrienne Russell, Mary Laird Wood Professor of Journalism and the Environment, University of Washington*

"This is a terrific book, deeply unsettling yet ultimately hopeful. Carefully argued, innovatively researched, and written with fierce optimism, it seeks to reframe and revision a prevailing understanding of climate change. It endeavours to add new and necessary voices to a conversation that has been lopsided and too often has ignored Indigenous input. Read it!"—*Mark Cronlund Anderson, Professor in the Department of History, University of Regina*

Media and Transnational Climate Justice

Simon Cottle
General Editor

Vol. 22

The Global Crises and the Media series
is part of the Peter Lang Media and Communication list.
Every volume is peer reviewed and meets
the highest quality standards for content and production.

PETER LANG
New York • Bern • Berlin
Brussels • Vienna • Oxford • Warsaw

Anna Roosvall and Matthew Tegelberg

Media and Transnational Climate Justice

Indigenous Activism and Climate Politics

PETER LANG
New York • Bern • Berlin
Brussels • Vienna • Oxford • Warsaw

Library of Congress Cataloging-in-Publication Data

Names: Roosvall, Anna, author. | Tegelberg, Matthew, author.
Title: Media and transnational climate justice: indigenous activism and climate politics / Anna Roosvall and Matthew Tegelberg.
Description: New York: Peter Lang, 2018.
Series: Global crises and the media; Vol. 22 | ISSN 1947-2587
Includes bibliographical references and index.
Identifiers: LCCN 2017022476 | ISBN 978-1-4331-3488-3 (hardback: alk. paper)
ISBN 978-1-4331-3487-6 (paperback: alk. paper)
ISBN 978-1-4331-4599-5 (ebook pdf)
ISBN 978-1-4331-4600-8 (epub) | ISBN 978-1-4331-4601-5 (mobi)
Subjects: LCSH: Indigenous peoples—Ecology. | Indigenous peoples—Politics and government. | Climatic changes—Political aspects. | Climatic changes—Moral and ethical aspects. | Environmental justice. | Climatic changes—Press coverage. | Mass media and the environment. | Conference of the Parties (United Nations Framework Convention on Climate Change).
Classification: LCC GF50 .R66 2018 | DDC 363.738/74—dc23
LC record available at https://lccn.loc.gov/2017022476
DOI 10.3726/b12949

Bibliographic information published by **Die Deutsche Nationalbibliothek.**
Die Deutsche Nationalbibliothek lists this publication in the "Deutsche Nationalbibliografie"; detailed bibliographic data are available on the Internet at http://dnb.d-nb.de/.

Cover photo: Shutterstock.
Cover artwork and retouch: Produktionsdesign i Stockholm AB

29 Broadway, 18th floor, New York, NY 10006
www.peterlang.com

TABLE OF CONTENTS

FIGURES

TABLES

ACKNOWLEDGMENTS

This book is the product of work done over many years, in many different places, leaving us with a long list of people to thank. We began working together in 2009 when the *MediaClimate* network, headed by Elisabeth Eide and Risto Kunelius, met for the first time in Istanbul, Turkey. The book would not have been possible without Elisabeth's and Risto's continuous, inexhaustible support. We are grateful for their invaluable intellectual input throughout the years, for the wonderful example they set as collaborative leaders, and for the opportunities they created for us to travel to UN climate summits as well as several relevant conferences and network events. We also wish to thank the other contributors to this network, researchers from more than 20 different countries (the exact number has varied slightly over the years). We are grateful to them for the data they contributed—which we draw on in this book—and for creating such a dynamic, fruitful environment at network gatherings. A special thanks goes to Dmitry Yagodin for helping us to access and process this network data.

We are grateful to several students and research assistants who helped us in various ways over the years. For coding of the Swedish press material, thanks to Maria Kabatanya (2009), Markus Mattisson (2011) and Emilia Roosvall (2013, 2015). At York University, Bryn Ludlow provided valuable research support prior to our fieldwork at the Paris climate summit (2015).

Kristina Stenström at Stockholm University helped to transcribe some of the Swedish interview material (2016). We are also tremendously grateful to Ragnar Dybeck for creating the book cover, and to Ragnar and Oscar Dybeck for creating some of our visual material.

Thanks to Karin Becker and Anna-Lill Drugge for generous sharing and helpful scholarly input and advice. For input on specific chapters we thank Lilie Chouliaraki, Florencia Enghel, Kate Nash, Alexa Robertson, Adrienne Russell and Andreas Widholm. Their comments have been invaluable to us, and have helped us immensely in clarifying our arguments. Any remaining issues are, of course, our own responsibility.

Anna wishes to thank the Department of Media and Communications at the London School of Economics and Political Science for hosting her as a Visiting Fellow during the autumn semester of 2016, when much of the writing for this book was done. Special thanks to Lilie Chouliaraki for hosting Anna, for providing invaluable commentary on chapter drafts, as well as for being welcoming, generous and constantly inspiring. Huge thanks to Nick Couldry as well for taking time to discuss the project at several meetings and for the welcoming spirit. His input and advice were very helpful for the project's development. Thanks also to Terhi Rantanen and Wendy Willems for pointing out fruitful paths to follow.

Matt wishes to thank his colleagues in the Department of Social Science at York University, especially Lisa Drummond and Kimberley White for providing invaluable feedback on the project at a critical juncture.

We are both grateful for thought-provoking questions and other input provided by our students, particularly those who took Media, Justice and Human Rights in the International Master Program in Media and Communication Studies at Stockholm University.

The project would not have been feasible without the funding received from Helsingin Sanomat Foundation; the Faculty of Liberal Arts and Professional Studies Minor Research Grant program at York University; and the Leading Research Environment in Global Media Studies and Politics of Mediated Communication in the Department of Media Studies at Stockholm University.

We also want to express our sincere gratitude to Series Editor Simon Cottle for his words of encouragement and insightful comments, and to the staff at Peter Lang; especially Acquisitions Editor Kathryn Harrison for her patience and continued support, and Production editor Jackie Pavlovic for facilitating the process.

The indigenous activists, community leaders and politicians we interviewed deserve the grandest thank you of all. This book would not exist if they had not generously shared their time, knowledge, insights, ideas, histories, experiences and perspectives. Thus we thank: Saoudata Aboubacrine, Marie-Josee Artist, Dmitry Berezhkov, Raymond de Chavez, Olga Atsynga Letykai Csonka, Justeene Dammay, Anthony Debbarma, Jon Petter Gintal, Lene Kielsen Holm, Áile Jávo, Curtis Konek, Jordan Konek, Vibeke Larsen, India Logan-Riley, Mikaele Maiava, Diel Mochire, Per Jonas Partapuoli, Francois Paulette, Tito Puanchir, Victoria Sharakhmatova, Jannie Staffansson, Teanau Tuiono and Allison Akootchook Warden. We also wish to thank NGO worker and interviewee Tiina Kurvits. Special thanks to interviewee Pirita Näkkäläjärvi, journalist, activist and indigenous media manager, for contributing concrete advice on a research presentation and helpful suggestions in general.

Last but not least we want to thank our families, for book related discussions and for much needed diversion; for being our inspiration and for all the fun. Thanks to Mårten, Julia, Emilia and Moltas, and honorary family member Joel, on Anna's team. And thanks to Lori, Elliott and Casey in Matt's gang.

Anna Roosvall and Matthew Tegelberg
Stockholm and Toronto

ABBREVIATIONS

APP	Ability to Pay Principle
BPP	Beneficiary Pays Principle
CHI	Common Human Identity
CHS	Common Human Status
COP	Conference of the Parties
COP 15	Copenhagen Climate Summit (2009)
COP 16	Cancun Climate Summit (2010)
COP 17	Durban Climate Summit (2011)
COP 19	Warsaw Climate Summit (2013)
COP 21	Paris Climate Summit (2015)
IPCC	Intergovernmental Panel on Climate Change
OECD	Organization for Economic Cooperation and Development
NGO	Non-Governmental Organization
PPP	Polluter Pays Principle
TEK	Traditional Ecological Knowledge
UN	United Nations
UNFCCC	United Nations Framework Convention on Climate Change

UNESCO	United Nations Educational, Scientific and Cultural Organization
UNDRIP	United Nations Declaration on the Rights of Indigenous Peoples
UDHR	Universal Declaration of Human Rights
UN ILO 169	Convention Concerning Indigenous and Tribal Peoples in Independent Countries

· 1 ·

INTRODUCTION

Calling for Climate Justice!

One of the most complex and crucial puzzles in a global age is how to handle issues of injustice, as they increasingly transgress the borders of nation-states (Fraser, 2008, 2014; Sen, 2009). Concurrently, one of the most pressing global crises of our times is the crisis of climate change (Cottle, 2009). These issues are intimately connected, as expressed in discussions surrounding who (i.e. what states) is responsible for causing climate change, and concerning who (i.e. what states) is responsible for taking action to mitigate the associated risks. Most people encounter such discussions in the media, if they encounter them at all. Others, who experience climate change concretely in their everyday lives, turn to activism as a means of affecting these discussions in politics as well as in the media. Increasingly, the voices we hear calling for climate justice are the voices of indigenous peoples, who often dwell—like climate change—in territories that transcend the boundaries and logics of nation-states. At the Paris climate summit in late 2015, we experienced a multiplicity of these compelling calls. They started before the summit began on the day of the Global Climate March:

Paris, 29 November 2015

Today the global climate march was supposed to take place in central Paris, just a few days before the annual UN climate summit would open at Le Bourget, in the suburbs nearby. But, a few weeks earlier the city fell victim to an atrocious terrorist attack. 130 people were killed and over 350 wounded by coordinated attacks in a restaurant, at a bar near Stade de France, on the streets, and during a rock concert at the Bataclan theatre. Consequently, a state of national emergency was declared in France and all public demonstrations were prohibited. Alternative commemorations and actions were announced nonetheless, some of which included indigenous activists.

We arrived at Place de la République in the morning. The Marianne monument at the centre was covered with flowers, messages, flags and candles to observe and pay respect to victims of the attacks. The widely reported shoe protest took place in the open space adjacent to the monument. It was arranged as a substitute for the global climate march, with shoes to symbolize all those who could not march. Gathered by Parisian activists and supporters, the shoes were sent by people from around the world, including Pope Francis and UN Secretary Ban Ki Moon (Jackson & Parry, 2015). Some shoes were decorated with messages, photos, balloons or flowers; some were simple and worn, others quite fancy. Together there were over 20,000 shoes, surrounded by placards stating "Paris on marche pour le climate!" and "Marche mondiale pour le climate", underlining the local ("Paris") as well as the global ("mondiale") aspects of the march.

At this point the manifestation was a silent one. We moved on to catch the tail end of an indigenous prayer ceremony, held in a park opposite to the Bataclan, a short walk from Place de la République. The fence surrounding the long narrow park was covered in drawings, including a widely circulated image of the Eiffel Tower merged with the peace sign. There were flowers, pages copied from newspapers, photos of people who had been killed, teddy bears, flags, letters and messages such as "Liberté, egalité, fraternité". The sidewalk was lined with candles.

Inside the park a family was playing table tennis at a ping-pong table located in a recreational area. As we approached the entrance, across from the Bataclan, we saw several young indigenous activists being interviewed by the media. They had just held a healing ceremony inside the playground to "stand in solidarity with the victims of the terrorist attacks and climate change" (Indigenous Environmental Network, 2015). Outside the fence, Allison Akootchook Warden, an Iñupiaq artist and activist from Alaska, was interviewed by one journalist after another. Dallas Goldtooth, a Dakota and Dine Nation activist and spokesperson for the Indigenous Environmental Network, was being interviewed inside the playground. He spoke about indigenous communities as frontline communities in the face of climate change, and underlined that indigenous peoples understand what it means to have great loss, such as the one experienced in connection with the terrorist attacks in Paris. Eagle Woman (Kandi Mossett) of the Mandan, Hidatsa and Arikara Nation in North Dakota, who held a prominent position at several protest actions, manifestations and cultural events during the conference, gave an emotional testimony about peace and hope as well as the fatal effects of fracking. She said, while crying, "... we're fighting for a systems change, and it will happen" (Indigenous Environmental Network, 2015).

Some interviews were produced for the Indigenous Environmental Network and distributed via their Facebook page. One such video from the healing ceremony included a statement indicating that the mainstream media had ignored the event (Indigenous Environmental Network, 2015). The video was later picked up by Canada's CBC News Indigenous, where it was embedded in a digital news story describing the actions of Canadian indigenous leaders at COP21 (Fontaine, 2015). In the playground, Kathy Jetnil Kijner, a young poet from the Marshall Islands was interviewed by a Norwegian journalist. She spoke about the severe situation in her low-lying native environment, at risk of flooding and eventually sinking into the ocean. After winning a local spoken word contest ("Spoken word for the world"), Kijner and other contest winners were sent to represent their communities in Paris. At the healing ceremony, they sang together with other indigenous activists before the interview session began. We managed to squeeze in a research interview with Allison Akootchook Warden, before she rushed off for a quick lunch with other activists so they could reach the Human Chain manifestation on time. Teanau Tuiono, a Māori activist from Ngāti Hine and Cook Islands who we would later interview, was also present at the healing ceremony, the Human Chain and many other protest actions and events organized by indigenous activists during the conference.

Like the indigenous activists and poets, we too moved on to the Human Chain manifestation, yet another expression of protest that occurred in spite of the declaration of Martial Law. It took place in Boulevard Voltaire, a short walk from the park adjacent to Bataclan where the healing ceremony was held. Indigenous activists, it was decided, would make up the first part of the chain, which then went on for two kilometres. Agence France-Presse reported that approximately 10,000 people took part (EWContributor, 2015; Jackson & Parry, 2015) with NGOs like 350.org and Action Aid making up stretches of the chain. In the indigenous section at the front, Teanau Tuiono stood together with Katy Jetnil Kijner and others holding a banner stating, "Stop fracking. Aotearoa. New Zealand. Our water is our future". Aotearoa is the Māori name for New Zealand. Other banners read "Honor Indigenous Treaties. Respectez les Traités Autochtones" and "We Speak Earth" (see Figure 1.1). Eagle Woman was leading the chanting: "When I say indigenous you say rights" (Eagle Woman: "Indigenous", Everyone: "Rights"), "What do we want? Justice! When do we want it? Now!", "When I say shut down you say fracking", and "When I say shut down you say coal." People from North America, South America and Oceania stood together chanting for climate justice and justice for indigenous peoples, which we shall see are two closely intertwined demands.

This account of our experience during the day of the global climate march captures an intriguing nexus of issues—*climate justice*, *geographical scales*, *media ecologies*, *activism* and *politics*—at a time when radical measures are increasingly demanded to address climate change. *Media and Transnational Climate Justice* examines this nexus through a trajectory seldom evoked by focusing on transnational rather than international aspects, and on colonized and

Figure 1.1. Banners from the indigenous section of the Human Chain manifestation.

marginalized indigenous peoples who do not have full representation at UN climate summits while at the same time being among those most exposed to injustice regarding climate change, politics, as well as in media coverage.

The purpose of this book is thus to examine and better understand the transnational dimensions of climate justice, particularly as they pertain to and are expressed by intra- and transnational indigenous activists and community leaders struggling to have their voices heard in predominately (inter) national political and media spheres. We ask what roles (different types of) media play—and could play to highlight and tackle issues of injustice—in their concurrent functions as agents and as arenas in relation to climate politics and climate justice activism. We argue, following Eide, Kunelius, and Kumpu (2010) and Eide and Kunelius (2012), that UN climate summits provide a rich case to study this problem since they constitute crucial defining, legitimating and mobilizing events when it comes to public understandings of climate change, as well as related political and communicative actions.

Media and Transnational Climate Justice scrutinizes political and ideological dimensions of the global phenomenon of climate change through interviews and observations with indigenous activists at UN climate summits. We combine this with extensive empirical research conducted on global media coverage of indigenous peoples and climate change in mainstream, social and indigenous media since 2007. Indigenous activists face particular challenges generating attention to their causes. For example, while they may get attention for their clothes, culture or spirituality, their political actions and ideological critiques, that increasingly take the form of demands for climate justice, have rarely been present in mainstream media coverage (Roosvall & Tegelberg, 2012, 2013, 2015). In this chapter, we introduce and develop four main themes that surfaced in our account of the day of the global climate march and which inform our analysis. Each of these themes constitute the subject of one chapter in this book: climate justice, geographical scales, media ecologies, and activism and politics. The concluding chapter forms a synthesis of the preceding ones by observing (dis)connections and a tension between the universal and the particular. We ultimately discuss transnational solidarity, not just as an action but also as a mode of communication applicable in politics as well as in the media.

From Climate Change to Climate Justice

Climate justice constitutes a specific way of understanding and relating to climate change; using the concept means seeing climate change as an issue that must be considered in relation to ethics, global equality, human rights, democratic accountability, issues of political participation and historical responsibility (see Chatterton, Featherstone, & Routledge, 2013; de Onís, 2012; Page, 2006; Shepard & Corbin-Mark, 2009; Shue, 2014). These issues are particularly relevant to demands for climate justice made by indigenous peoples in light of the historical legacy of colonization, and their ongoing marginalization from dominant political and media spheres (Alia, 2010; Anderson & Robertson, 2011; Meadows, 2001; Roosvall & Tegelberg, 2015). Before elaborating on climate justice, we must detail some general views and facts regarding climate change as such.

Hulme (2009, p. xxxix) defines climate change as "past, present or future change in climate, with the implication that the predominant—but not exclusive—cause of this change is human in origin".[1] It has long been clear that the emission of carbon into the atmosphere, from the burning of fossil fuels, is what causes dangerous climate change by blanketing the planet and keeping additional heat on the earth's surface. Thus climate change, as well as climate justice, discussions have largely concerned issues of how to regulate or end carbon emissions (Shue, 2014, p. 1). Such discussions reach larger publics through the media, thus pointing to the necessity of studying media output as well as attitudes towards the media, as we do in this book. Accounts of when exactly we came to know about climate change differ slightly. Some argue such knowledge existed in the nineteenth century, while others trace its origins to the 1950s (Brülde & Duus-Otterström, 2015, pp. 90–91). The first major study of global environmental issues, *The Limits to Growth*, was published by The Club of Rome think tank in 1972 (Hulme, 2009, p. 260; Urry, 2011, p. 165). It is however perhaps most common to draw the line in 1990, when the IPCC First Assessment Report was published, with the expressed aim of authoritatively informing the world's states of climate change research results (Brülde & Duus-Otterström, 2015, pp. 90–91).

Among the scientific community, evidence that global climate change is anthropogenic is now largely uncontested (Reese, 2016, p. 522). What is central to climate futures is thus human behaviour (Urry, 2011, p. 155). Relating to this, and to the fact that since 1900 more energy has been deployed than in all of prior human history (John McNeill, cited in Urry, 2011, p. 155), geologists have introduced a new epoch that embodies our current geological age: the Anthropocene. The term refers to an epoch, beginning with the industrial revolution and intensifying in the twentieth century, when human activity became the dominant influence on planetary geology, climate and ecosystems. In light of these developments, social scientists have increasingly begun to engage critically with questions concerning the cultural and political meaning of the Anthropocene (Lövbrand et al., 2015). We point to the necessity of studying cultural and political aspects of not only the Anthropocene at large, but also of climate justice in particular. We do so by analyzing media framings of indigenous peoples in cultural and political contexts.

Climate change today poses enormous challenges for humanity. Beyond the obvious effects of emissions on nature, climate change can affect health and hamper psychological well-being (Reser, Morissey, & Ellul, 2011; Swim et al., 2009), as testimonials from the indigenous activists we interviewed serve to detail. Negative health effects may in turn encumber engagement in nature conservation behaviour (Reese, 2016, p. 522), producing a negative downward spiral. This however was not the approach of the activists we interviewed, who were instead fighting for themselves and on behalf of those who could not; actions resembling the manifestation described above where shoes filled in for those who were unable to march. Climate change is thus not only transforming the physical world, it is also affecting the social realm (Hulme, 2009, p. xxviii). Since the 1970s, climate change has transformed from an object mainly of scientific interest to a subject of daily public discourse worldwide (Hulme, 2009, p. xxix). This public discourse largely takes place in the media (Boyce & Lewis, 2009; Boykoff, 2011; Doyle, 2011). The "story" of climate change is accordingly one of nature *and* culture (Hulme, 2009, p. xxviii). Humans are crucial actors in both of these spheres, and continuously re-create them (Hulme, 2009, p. xxviii). To a large extent, this re-creation takes place in the media.

Urry's (2011, p. 155) perspective on climate change puts society in focus as both the solution and the problem. In terms of the solution, according to Klein (2014, p. 409), climate change could be the basis of a "Marshall Plan for the Earth", helping us to tackle inequality. Hulme states, likewise, that "… the stories we tell about climate change and those we tell about sustainable development are inextricably linked" (2009, p. 280). He adds that "… a better understanding of climate variability and change could help alleviate human suffering" (Hulme, 2009, p. xxx), which we approach by detailing climate testimonies of interviewees (Chapter 3) and sketching a solidarian approach to representation in politics and media (Chapter 6). Conversely, climate change can exacerbate resource conflicts and forced migration as well as scapegoating and nationalism (Burke, Miguel, Satyanath, Dykena, & Lobell, 2009; Schubert & Schnellhuber, 2009; see also Reese, 2016, p. 522), which we relate to in our theoretical understanding of climate justice (Chapter 2) and in our discussion of activism (Chapter 5).

According to Hulme (2009, p. 361), the idea of climate change should be used to take a step back to rethink our broader social goals, particularly why we live on this planet at all. We argue that talking about climate change in terms of climate justice is a crucial way of doing this. Indeed, the idea of climate change can provoke new ethical thinking concerning our relationship with our future (Hulme, 2009, p. 363). Like environmental justice, climate justice focuses on humans rather than the possible extinction of certain species of animals (Harvey, 1996, pp. 286, 399; Pezzullo & Sandler, 2007, pp. 8–9). Climate change does not however affect all humans to the same degree nor in the same sense. Often those exposed to the most severe impacts have done the least to create them (Johnson, 2009). This is certainly relevant for many indigenous groups living in sustainable ways close to nature (see also Tauli-Corpuz, 2009).

Justice issues are also often somewhat paradoxically viewed as obstacles to tackling climate change, in that nations may halt actions by claiming others must act first, or that nothing can be done until *everyone* acts (Reese, 2016). Others argue that a justice perspective can help tackle climate change beyond climate justice. The more people see the imbalanced distribution of climate change risk as unjust, the more they are likely to: (a) engage in climate-protective behaviour; (b) commit to motivating others to behave climate-protectively; and (c) express commitment to climate protective laws (Reese, 2016, p. 524, see also Kals & Russell, 2001). Attention to justice in relation to the climate will also open up spaces for dialogue between people who support diverse policies and governing structures (Reese, 2016, p. 524; Sikor, Martin, Fisher, & He, 2014). As we shall argue, particularly in Chapter 5 on activism and "the political", such dialogue is crucial not only for climate justice, but also for democracy at large. Moreover, one of Reese's (2016, p. 524) hypotheses in his study on a Common Human Identity as a pathway to global climate justice reads: "A common human identity elicits an inclusive representation of the human world, resulting in environmental justice beliefs considering all individuals and social groups". The emphasis on *representation* here again points to the relevance of media and communication studies in relation to climate justice.

While there is a "concern for the disadvantaged" in media coverage on climate justice (Schmidt & Schäfer, 2015, p. 546), we argue it remains crucial to pinpoint *who* is considered disadvantaged and on what grounds. Who, in other words, is the subject of injustice (Fraser, 2008)? Is it national citizens? Intra- or transnational groups, such as indigenous peoples? We set the stage for discussion of the complex issues of determining the who, the what and the how of justice in Chapter 2, and return to different aspects of this problem in the subsequent chapters. We wish to underline in this regard particularly that places directly exposed to devastating climate change effects are often inhabited by indigenous peoples, with livelihoods and worldviews closely connected to the natural environment. The complex ecosystems these groups inhabit, in the Arctic, the Amazon, the Andes, and the Australian flatlands, to name a few, are geopolitically "remote" in relation to centres of power in media and in politics. A situation epitomizing Johnson's contention (2009, p. 297) that "Those who feel the most severe impacts [of climate change] may be in the worst position to address them".

Geographical Scales

On 23 September 2013, the High Level Advisory Committee to the Climate Justice Dialogue, an initiative of the Mary Robinson Foundation—Climate Justice and the World Resources Institute, issued a *Declaration on Climate Justice* at the UN (see Shue, 2014, p. 340–342). The declaration underlines that not only nations, but also peoples should be treated fairly in efforts to tackle climate change. It thereby addresses climate justice as an international as well as intra- and *transnational* issue; one that encompasses all of humanity, globally, while at the same time stressing the importance of local initiatives. The Mary Robinson Foundation declaration makes one mention of indigenous peoples: "For many communities, including indigenous peoples around the world, adaptation to climate change is an urgent priority that has to be addressed much more assertively than before" (World Resources Institute, 2013).

The balance this declaration aims to strike between multiple geographical scales and subjective experiences has not characterized climate governance initiatives to date. While climate change is frequently defined as a *global* issue, at UN climate summits negotiations take place between *nations* in an *international* setting. In this context, indigenous peoples occupy an ambiguous position. They often constitute *transnational* or *intra-national* communities, sometimes recognized as nations with partial independence, but rarely granted the right to negotiate on par with nation-states in international political settings. At COP summits, in other words, nations negotiate while indigenous peoples and other transnational groups often end up on the margins or outside of official negotiation processes. We discuss these issues further by describing the geopolitical cartography of UN climate summits (Chapter 3), and by detailing the struggle to be heard and seen in the media as well as in the political processes that govern climate summits (Chapters 4 and 5). Major media reporting from summits is predominately *national* (newspapers, broadcasters, etc.) and tends to marginalize voices that do not conform to these structures.

Rao writes that "Globalization defines our era. It is what happens when the movement of people, goods, or ideas among centuries and regions accelerates" (2015, p. 121). The origin of the UN climate summits can be seen as parts of these accelerations. Globalization is sometimes divided into diverse aspects: technological, social, cultural, economic and political (Appadurai, 1996; Rantanen, 2005). We understand globalization to encompass all these aspects, and view it as "a process in which worldwide economic, political, cultural and social relations have become increasingly mediated across time and space" (Rantanen, 2005, p. 8). Thereby media plays a crucial role in globalization by linking together diverse geographical scales. Globalization can more specifically consist of international as well as transnational relations. International relations are signified by interaction between nation-states. We define the transnational, by contrast, as a stretching out *across* two or more states, which applies to indigenous peoples like the Sami (see Plaut, 2017).

The nation-state is generally viewed as the most central actor in efforts to tackle climate change (Brülde & Duus-Otterström, 2015, p. 84). States have a strong moral obligation to work toward shifting away from fossil fuels to clean energy sources that do not contribute to global warming (Brülde & Duus-Otterström, 2015, p. 82). But questions concerning responsibility for climate justice are linked to diverse geographical scales. Who is the subject in climate justice arguments (Schmidt & Schäfer, 2015, p. 536)? Is it citizens of the nation-state that are the target audience of a particular media outlet? Or does the category apply more broadly to citizens of all nations? And, finally, can groups with transnational identities, such as indigenous peoples, be included in the constitution of these subjects, both in terms of the way they are organized in politics and expressed in the media?

These challenges call for a *multi-spatial* approach to the study of media and identities (Georgiou, 2011, p. 211). Here we apply such an approach by considering multiple spatial scales in media representations of indigenous peoples. Focus on transnational scales of climate justice has been rare in media research, as has attention to indigenous peoples. This focus is crucial for understanding issues of transnational injustice more generally, for instance concerning other pressing global crises such as migration. It can also be argued that climate governance as such must be multiscale; it cannot be managed through the nation-state alone, or even primarily (Hulme, 2009, p. 314; Prince & Rayner, 2007). Societies vary in vulnerability as well as in their capacity for resilience (Giddens, 2009, pp. 176–177; Urry, 2011, p. 166). There are also *intranational* variations within societies, between diverse parts of nation-states. These can lead to more harshly felt climate change effects among indigenous communities, who sometimes have livelihoods that are dependent on nature and animals, and strong connections to their lands.

Carbaugh and Cerulli (2013, p. 6) explain that "Places—those we visit and dwell within briefly, those we call home, and those from which we are sometimes displaced—serve us in particular ways as geography for our thinking, gathering our thoughts, holding our attentiveness." The formulation of this "geography for our thinking" partly occurs through news reporting (Corbett, 2012; Robertson, 2015; Roosvall, 2010). Christensen (2013, p. 27) found in recent years *scalar transcendence*, a mixing between local and global scales, has increased in the mediated geography of our thinking regarding climate change. In Chapters 3 and 4 we consider whether these findings pertain to our research material and distinguish between natural and mediated experiences of climate change. In the next section we discuss media ecologies further.

Climate News Media Ecologies

The deliberations and civil society actions that take place at UN climate summits are made public by media. Media, in fact, make a huge contribution to the structuring of these summits, which can be defined as media events on a supranational level (Dayan & Katz, 1992; Eide, Kunelius, & Philips, 2008, Eide et al., 2010). During these events, media take on the role of actors that can shape public perceptions and opinions by *framing* debates pertaining to the summits and to climate change more broadly (Hulme, 2009). Hence, it is imperative to examine semiotic and material dimensions of the news media ecologies that form around and shape perception of the climate issues debated and contested at UN COPs. We contribute to this effort, in the tradition of media ecology theory (Logan, 2007; McLuhan, 1964; Postman, 2000),[2] by exploring relationships between indigenous activists and the media in which they are represented and engaged with during COP summits. But first, here we briefly consider some relevant findings and debates in a growing body of literature on climate change communication.

Climate change can be framed in multiple ways; for instance as an ecological issue, a political-economic, cultural and social issue, a scientific issue (see Christensen, 2013, pp. 38–39) and necessarily, as we wish to highlight, as an issue of justice. Framing occurs to a significant degree in the media. Media are important agenda-setters in the climate change debate, specifically in the formation of domestic opinions (Boykoff & Boykoff, 2004; Schmidt & Schäfer, 2015, p. 537). Media institutions have *power* to "steer" political debate and policy decisions (Christensen, 2013, p. 27). Since climate change is now recognized as an agenda-setting issue for national and global policy, print media have been identified as research objects due to their crucial role as "intermedia agenda-setters" (Reese & Danielian, 1989; Schmidt & Schäfer, 2015, p. 537). They are concurrently the providers of prime information about climate change for people (Sampei & Aoyagi-Usui, 2009, see also Schmidt & Schäfer 2015, pp. 537–538). Others have focused on the role digital and social media platforms have increasingly come to play in shaping public knowledge of climate change (Painter et al., 2017; Williams, McMurray, Kurtz, & Lambert, 2015). Print media thus constitutes a significant part of our empirical material. It is accompanied by Twitter data, sample articles and video from indigenous-produced or co-produced material on diverse social media platforms. In our interviews with indigenous activists, we also asked about their views on and use of diverse media, specifically concerning news.

Research has shown that national domestication is a common feature of media reporting on climate change around the globe (Eide et al., 2010; Olausson, 2014; Schmidt & Schäfer, 2015). Schmidt and Schäfer demonstrated this by examining climate change discourses in German media, and in a comparative study of US, Indian and German media. German media tended to represent Germany as a forerunner in climate change policy and discourse. Similar tendencies were found in a study of editorials from high income countries, specifically in sample Japanese and Swedish editorials (Roosvall, 2016). National domestication is bound to leave indigenous peoples out, as they are often situated beyond nation-state frames. Hence, we connect media framing to the concept of framing in justice theorizing (Fraser, 2008). Thus, when considering media framing it is essential to identify not only what is being framed and how, but also who or what is *not* being framed or being *misframed* (Fraser, 2008). Misframing is understood here as a form of misrepresentation, where just political representation is hindered by community boundaries which prohibit some people from the chance to participate (Fraser, 2008, p. 19). Misframing arises for instance "… when the state-territorial frame is imposed on transnational sources of injustice" (Fraser, 2008, p. 114). We introduce this trajectory in Chapter 2 and elaborate on how it plays out in diverse contexts throughout the book.

We investigate tensions and (dis)connections between mainstream print media and digital communication. "The underlying point of … media studies," Couldry writes (2012, p. 210), "is to study how media contribute to the conditions of knowledge and agency *in the world* and so to understand better whether media enable us to live well together". Thus, with Chouliaraki (2008), we ask: what sense of responsibility do media representations in diverse media outlets cultivate regarding (distant) others? Can the spectacle of suffering that climate change reporting sometimes stages lead to constructive forms of public action toward (distant) others? These questions are addressed in Chapter 5, in particular, where we focus on the role of activism and the ways media witnessing is performed across different media forms.

Activism and Politics

Climate politics is largely carried out on the international and intranational levels, between or within states. Therefore, transnational groups tend to fall outside of this scope and have had relatively scarce opportunities to have an impact. Transnational groups are instead often found on the more ideological side of climate politics, performing "the political" (Mouffe, 2005, 2013) through activism (Carvalho, 2005; Uldham & Askanius, 2013).

The importance of activism in general, and environmental/climate activism in particular, has attracted increasing interest in media studies. We argue that indigenous activism has not yet been given enough attention despite becoming an increasingly impactful force. Moreover, it provides an excellent case for understanding transnationalism as well as the complexity of justice and geographical scales in a globalizing age. Here we briefly introduce significant parts of the recent history of environmental and climate justice activism, particularly in relation to the media, before delving into our own material.

In the 1970s, the environmental justice movement emerged in the United States with a series of protests and demonstrations, by people of colour and the poor, opposing the zoning of hazardous waste landfill sites and other environmental threats (Bullard, 1990). The movement was initially connected to local issues and grassroots actors, but expanded during the 1980s eventually culminating in the First National People of Color Environmental Leadership Summit in 1991. Its main aim was to reveal environmental racism: practices and policies that were disproportionately affecting the environment in communities of colour (Bullard, 1990). This later became the catalyst for research investigating uneven concentrations of environmental threats in poor, racialized communities in many nations (Agyeman, Bullard, & Evans, 2003). Endres (2009) has noted that discursive aspects of such environmental colonialism require further study. We analyze precisely such aspects of environmental and, more specifically, climate threats, through our focus on the media as framing agents, and as arenas for frames produced by others.

While human health has always been a concern for mainstream environmentalists, the environmental justice movement placed *social inequalities* at the top of its agenda (Pezzullo & Sandler, 2007). By doing so, it moved from nature conservation and the survival of endangered species toward human survival, particularly concerning poor, marginalized peoples identified as "other" (Harvey, 1996, pp. 286, 399; Pezzullo & Sandler, 2007, pp. 8–9). The movement redefined the environment to encompass the totality of life conditions: air, water, wages, housing, education, health, equity, and justice, in connection with the need to end discrimination (Harvey, 1996, 391; Szasz, 1994, 151). How do news media relate to these issues? Are they perhaps more concerned with polar bears than with indigenous people living close to polar bears? We discuss these questions further in Chapter 4.

Scholars and activists describe the devastating impact environmentally damaging politics have had on local environments and ways of life (Agyeman, Cole, Haluza-DeLay, & O'Riley, 2009; LaDuke, 1994, 1999; Westra, 2007). More specifically, these studies have shown that indigenous groups are often at the centre of environmental injustice (Agyeman et al., 2009, p. 3). This research trajectory often focuses on demands for justice at the local level. Harvey notes, however, that all environmental movements fighting for particular causes have a choice to continue struggling locally or strive for a more universal politics, which "[…] involves a move from one level of abstraction—attached to place—to quite different levels of abstraction capable of reaching across space in which communities could not be known in the same unmediated ways" (Harvey 1996, p. 399). Thus, by necessity, moving from the local to the global entails turning to the media. This calls for further investigation of the media's role in environmental and climate injustice, and more particularly to peoples exposed to these injustices, such as the indigenous groups placed at the centre of this study.

Media studies of social activism have gradually started merging with media studies of environmental activism, especially concerning digital media (Askanius & Uldham, 2011; Davidsen, 2011; Gavin & Marshall, 2011; Segerberg & Bennett, 2011; see also Roosvall & Tegelberg, 2015). Others place environmental activism in the context of political activism and the media (Christensen, 2011; Cottle & Lester, 2011) or in the context of social justice (Sowards, 2012). Recently, a shift from *collective action*, including the reinforcement of group identity and the development of shared frames, towards *connective action*, which entails a more personalized engagement within loose networks, has been noted as a key feature of the increasing intersection of social movements and digital communication (Bennett & Segerberg, 2011; cf. Thorsen, Edgerly, Kliger-Vilenchik, Xu, & Wang, 2016). The *organizational capacities* of digital communication are underlined in the latter approach, where the importance of group identity diminishes (Bennett & Segerberg, 2012, 2013).

Zooming in on the climate justice movement more particularly, some claim it emerged forcefully at the 2009 climate summit in Copenhagen (Chatterton et al., 2013; Ytterstad & Russell, 2012), while others trace its origin back to the year 2000 (de Onís, 2012). Ytterstad and Russell (2012, p. 247) study "… the role of journalism in mediating the political goals and arguments of climate activists". Like Askanius and Uldham (2011), they focus on the mediation of climate justice at UN climate summits. These studies make crucial contributions to understanding activism, media and climate justice in the context of the global justice movement. We contribute to these discussions and wish to develop them further by focusing on how scales of justice relate to geographical scales, particularly in relation to indigenous peoples, and how these groups relate to collective and/or connective action.

Transnational civil society has an important role to play at the UN climate summits, especially outside of negotiations between nation-states (Hulme 2009, p. 305). Carvalho (2007, cf. Hulme, 2009, p. 307) has for instance shown that a "transformative global citizenship" has been constructed by Carbon Reduction Action Groups (CRAGS) rooted in ideas of ecological citizenship. They have sought to "transform the relations between political power, consumption and global responsibility" (Carvalho, 2007, cf. Hulme, 2009, p. 307). This corresponds to claims in the Declaration on Climate Justice: "As part of global collective action, greater emphasis should be given to the role of diverse coalitions that are already emerging at the community, local, city, corporate and country levels, and the vital role they play in mobilizing action" (World Resources Institute, 2013; see also Shue, 2014, p. 342). Indigenous peoples work like this on a transnational level, while being rooted in diverse local and national contexts. We elaborate on what this means for discussions of climate justice and what the media can learn from this in Chapters 5 and 6.

Mixed Methodological Approach

The material for this book has been collected over many years, and several methods have been used; sometimes in combination, sometimes separately. The quantitative content analysis (Bryman, 1996) centres on a longitudinal sample of print media coverage of UN climate summits (COPs) collected by the MediaClimate network between 2007 and 2015.[3] In this book we mainly refer to material collected during COP15 in Copenhagen (2009), COP17 in Durban (2011), COP19 in Warsaw (2013) and COP21 in Paris (2015). This timeframe coincides with significant changes in the news media landscape, particularly concerning the growing influence of social media. Our analysis of print media primarily focuses on mainstream print journalism in Sweden and Canada.[4] The changing media landscape also informed our decision to examine the representation and practices of indigenous actors in digital media networks. This includes the analysis of a small sample of Twitter data using the hashtag #indigenousCOP21, as well as other digital media material, produced by indigenous peoples and, in some cases, by their collaborators. In addition, we discuss the practices of official state indigenous media, drawing on the example of Sami broadcasting in Finland.

Qualitative research methodologies are also applied to several types of material. We use framing analysis (Altheide, 1997; Entman, 1993, 2007) and discourse theoretical analysis (Laclau & Mouffe, 1985; Winter-Jørgensen & Phillips, 2002) to study media representations of indigenous peoples. The discourse theoretical analysis we employ is sometimes operationalized together with central theoretical concepts in order to scrutinize references to and attitudes towards redistribution, recognition (of identity versus status) and representation (political participation) as remedies for injustice (Fraser, 2000, 2008, see Chapter 2). For example, in Chapter 5 we conduct analysis of media witnessing (see Frosh & Pinchevski, 2009; Peters, 2001) by drawing on this combination of qualitative method and theoretical underpinnings.

Our interview research is guided by Bryman's (1996, p. 46) position that open-ended interviews help to facilitate the analysis of social issues from the perspective of the actors involved. In each chapter, interviews provide supporting evidence of the analytical pertinence of a theoretical framework. Moreover, they open up spaces of representation and reflection on transnational indigenous activists' views concerning the politics and mediation of climate change. Finally, they inform our understanding of transnational climate justice and the diverse roles of the media. The interview transcripts were initially organized according to themes, roughly corresponding to the themes of the chapters in this book. In this endeavour we were, in accordance with discourse theoretical analysis, attentive to the use of concepts and how they are related to each other (Laclau & Mouffe, 1985; Winter-Jørgensen & Phillips, 2002; see also Krzyżanowski and Forchtner, 2016, on the importance of concepts in critical discourse studies). We also scrutinized the transcripts in search of concepts that transcended the identified themes (e.g. recognition).

By "indigenous activists", we refer to the indigenous actors we interviewed at UN climate summits in Durban (2011) and Paris (2015). They attended these summits to represent their communities at climate demonstrations and, in some instances, as participants with varying degrees of influence in the ongoing political deliberations.[5] Hence, in the interview study our scope is limited to indigenous actors who function in this particular role (with a few exceptions, see Table 1.1, and see the additional interview with a Sami journalist). Tables 1.1 and 1.2 illustrate the often transnational identities of the organizations and constituencies these activists represent; frequently communities with vast territories spanning across several nations or continents (see especially columns 3 and 4).

"Indigenous peoples" is a broad term used to refer to groups that "… retain social, cultural, economic and political characteristics that are distinct from those of the dominant societies in which they live" (United Nations, n.d.). To account for the diversity of indigenous peoples, the UN Permanent Forum on Indigenous Issues has established a broad system of recognition that encompasses the following: self-identification as indigenous peoples; maintenance of historical continuity with pre-colonial societies; strong ties to traditional territories and natural resources; distinctive social, political and economic systems; distinct languages, cultures and beliefs; and non-dominant groups within the society (United Nations, n.d.). When recognized as "indigenous" a group is sometimes ascribed particular status within a nation. Indigenous peoples' rights are stipulated in and regulated by universal and particular declarations, covenants, conventions (mainly originating from diverse UN bodies) and treaties (originating at the national and region levels). We relate to some of these explicitly in Chapter 6. Finally, we pluralize indigenous peoples so as to underline differences between indigenous groups (Smith, 1999).

We interviewed a total of 25 people for this study. In 2011 and 2015, we interviewed 23 indigenous activists and leaders and 1 non-indigenous NGO staffer (Tiina Kurvits). 7 of these interviews were conducted in December 2011 at COP17 in Durban, South Africa. 17 were conducted in November and December 2015 at COP21 in Paris, France. Sami journalist and head of Sami broadcasting in Finland Pirita Näkkäläjärvi responded to interview questions via email in January 2017.[6] Each of the interviews has been transcribed verbatim and we were granted permission to publish the names of the interviewees.[7]

Tables 1.1 and 1.2 provide background information on each of the interviewees. This includes their organizational affiliation, indigenous group and geographical region. Table 1.2 also identifies the varying degrees of access interviewees had to the COP process. The indigenous activists we spoke with fell into one of four categories: (a) Delegate with "Party" status (access to COP decision-making arenas, see Chapter 3), (b) Delegate with "Observer" status (access to some COP sessions, exhibitions and side-events, see Chapter 3), (c) Institutional support (NGO/Philanthropy/Academia), and (d) Non-accredited, grassroots actors who are either self-supported or crowd funded. We adhere to the way interviewees identified themselves, their affiliation and their access. In addition, we point out the region and nation-state they dwell in since this is an important dimension of the political process at the COPs. The regions are specified according to the way they were distinguished in the Indigenous Peoples' Pavilion at COP21 in Paris. Sometimes regions may overlap, as in the cases of the Arctic, North America and Russia.

Table 1.1: Summary of COP17 interviewees by region and organization.

Interviewee	Region	People/Nation-state	Organization
Curtis Konek	Arctic/North America	Inuit, Arviat, Nunavut, Canada	Inuit Youth Delegation
Jordan Konek	Arctic/North America	Inuit, Arviat, Nunavut, Canada	Inuit Youth Delegation
Vibeke Larsen	Arctic	Sami, Samiland, Norway	Sami Parliament, Member of Norwegian Delegation
Tiina Kurvits	Arctic/North America	Non-indigenous, Ottawa, Canada	Many Strong Voices
Raymond de Chavez	Asia	Igorot people, Philippines	Tebtebba Foundation, Indigenous Peoples' Centre for Policy Research and Education
Francois Paulette	North America	Dene, Northwest Territories, Canada	Indigenous Environmental Network
Tito Puanchir	Latin America and the Caribbean	Shuar, Ecuador	Confederation of Indigenous Nationalities of the Ecuadorian Amazon (CONFENIAE)

Table 1.2: Summary of COP21 interviewees by region, organization and summit access.

Interviewee	Region	People/Nation-state	Organization	Access	Status
Áile Jávo	Arctic	Sami, Samiland, Norway	Sami Council President	Yes	Observer
Per Jonas Partapuoli	Arctic	Sami, Samiland, Sweden	Sami Council, International Center for Reindeer Husbandry	Yes	Observer
Lene Kielsen Holm	Arctic	Inuit, Greenland	Inuit Circumpolar Council (ICC)	Yes	Observer
Jon Petter Gintal	Arctic	Lule Sami, Samiland, Norway	Sami Parliament, Senior Advisor, Department of Rights and International Affairs	Yes	Party, Norway

Interviewee	Region	People/ Nation-state	Organization	Access	Status
Jannie Staffansson	Arctic	Sami, Samiland, Sweden	Sami Council, Consultant on environmental politics and environmental issues	Yes	Observer
Olga Atsynga Letykai Csonka	Arctic/ Russia and Eastern Europe	Chukchi people, Western Chukotka, Russia	No organization, made several throat singing performances	No	No access
Dmitry Berezhkov	Arctic/ Russia and Eastern Europe	Itelmen people, Kamchatka, Russia	Asia Indigenous Peoples Pact Foundation	Yes	Observer
Victoria Sharakhmatova	Arctic/ Russia and Eastern Europe	Itelmen people, Kamchatka, Russia	Indigenous People of Kamchatka Krai	Yes	Observer
Allison Akootchook Warden	Arctic/ North America	Iñupiaq people, Kaktovik, Alaska, US	Sierra Club & Indigenous Environmental Network	No	No access
Justeene Dammay	Asia	Tuwali people, Ifugao Cordillera, Luzon Island, Philippines	Asia Indigenous Peoples Pact Foundation	Yes	Observer
Anthony Debbarma	Asia	Borok people, India	Asia Indigenous Peoples Pact Foundation	Yes	Observer
Saoudata Aboubacrine	Africa	Kel Tamasheq (also known as Tuareg people), Burkina Faso & Mali	Tuareg Women's Organization, Tin Hinan; African Indigenous Peoples Network (IPAN)	Yes	Party, Burkina Faso

Interviewee	Region	People/ Nation-state	Organization	Access	Status
Diel Mochire	Africa	"Pygmies"[8] of North Kivu, Democratic Republic of Congo	Network of Indigenous and Local Communities for the Sustainable Management of Forest Ecosystems in Central Africa (REPALEAC)	No	No access
India Logan-Riley	Pacific	Māori, Ngāti Kahungunu Heretaunga, New Zealand	Aotearoa Youth Leadership Institute, self-funded	Yes	Observer
Mikaele Maiava	Pacific	Tokelauan, Tokelau	Pacific Indigenous Network	Yes	Party, Samoa
Teanau Tuiono	Pacific	Māori, Ngāi Takoto, Ngāpuhi, Ngāti Hine, New Zealand and Cook Islands	Ngāti Hine, Māori organization	Yes	Party
Marie-Josee Artist	Latin America and the Caribbean	Kalina people, Biki Poika, Suriname	Association of Indigenous Village Leaders in Suriname	No	No access

Finally, we engaged in participant observation (Iorio, 2011, p. 13) of media events arranged by indigenous activists and representatives. At both summits, we attended press conferences or media events organized by indigenous organizations. Some events took place inside the summit halls in spaces designated for civil society activity. Others took place outside the summit in activist spaces around the city. At COP17 in Durban (2011) we were accredited and had access to the official summit halls. At COP21 in Paris (2015) we were not accredited and only had access to the Climate Generations Space, where NGOs had booths. This was also where the Indigenous Peoples' Pavilion and all its associated events were situated.

Studying Representations with the Represented

In *Decolonizing Methodologies: Research and Indigenous Peoples* (1999), Māori scholar Linda Tuhiwai Smith delivers a powerful critique of Western scholarly research, and makes a forceful call for its decolonization (see also Drugge, 2016; Ledman, 2015). The book exposes the complicity of Western research in "… a deep structure [that] regulates and legitimizes imperial practices" while dismissing indigenous worldviews and experiences as "… not fully human … not civilized enough to have systems" (Smith, 1999, p. 28). It then counters this dominant paradigm by articulating an indigenous research programme that includes 25 different projects (e.g. "Connecting," "Reframing," "Sharing", etc.). Heeding Smith's call to decolonize Western research, we take listening and learning from indigenous peoples as a starting point to map and critique their exclusion from dominant paradigms in climate change research, politics and communication.

In our interviews at COP17 we asked indigenous interviewees for their opinions on what research they thought was needed regarding media, climate change and indigenous peoples, so as not to overlook something from our own (non-indigenous) point of view. Several interviewees stressed the need for a focus on political aspects, and why such aspects are rare in media reporting on indigenous peoples (see Roosvall & Tegelberg, 2015). This critical insight informed our research going forward. In this book we discuss for instance the relationship between "politics" and "the political" (Mouffe, 2005, 2013) and we are particularly attentive to how political perspectives are represented in media output on climate change and indigenous peoples (see especially Chapters 4 and 5).

While we are not studying indigenous peoples as such, but rather *their representation in and relationships with the media*, we must still acknowledge our own positionality as Western researchers; specifically with a view to colonialism and the sustained power relations attached to it. We are white non-colonized researchers and thus enjoy privileged positions in terms of ethnicity, nationality, social class and education. Initially, we came to study media reporting on climate change and indigenous peoples because we noted a deficit of such reporting in the national print media in our home countries, Sweden and Canada, which are also both home to populations of indigenous peoples. In our field research at the COPs, observing the thousands of journalists and researchers who attend these events further compelled us to ask why so little attention was being paid to indigenous peoples. Throughout the research process, we have shared our work with indigenous peoples and asked them for input and feedback on what we should take into consideration. We realize we still cannot provide direct perspectives from the inside, which are obviously needed when it comes to indigenous peoples, climate change and the media. However, we do strive, to the extent possible, to provide perspectives from the inside *indirectly* through quoting the indigenous activists we interviewed.

The Structure of the Book

Following this introductory chapter, the book consists of five chapters which in turn: theorize climate justice in dialogue with input from the interviewees; detail the geographies of climate change and of UN climate summits through our observations and the accounts of interviewees; zoom in on media ecologies and media content; discussing activism, agency and voice in the context of agonistic democracy theory; and finally sum up the (dis)connections we found while also discussing opportunities for transnational solidarity as a mode of communication.

Chapter 2, "What is climate justice? Justice, climate and the media," theorizes climate justice by discussing broader theories on justice and responsibility, relating them to the media. The importance of re-theorizing justice in a globalizing age leads us to discuss climate justice along diverse geographical scales: international, intranational, transnational and global. Thus far much of the literature on climate justice understands the concept as having to do with justice between nation-states (Schmidt & Schäfer, 2015; Shue, 2014), largely focusing on which nations should pay (the most) for the necessary measures: the polluter, the beneficiary, or the one with the ability to pay (Brülde & Duus-Otterström, 2015). We add to these discussions by underlining transnational and intranational dimensions of climate justice, which are relevant for indigenous peoples who are often transnational (e.g. the Sami, the Inuit) and simultaneously marginalized within their respective nation-states (Roosvall, 2016; Roosvall & Tegelberg, 2015). We discuss furthermore if a sense of a Common Human Identity (CHI) could help implement climate justice globally (Reese, 2016). Definitions of climate justice posited by the interviewees, and further comments on its meaning in relation to responsibility, geography and the media, permeate the chapter. Thereby input from the indigenous activists we interviewed helps to pinpoint ruptures and deficiencies in previous understandings, and point to the need for more emphasis on the interconnected intra- and transnational scales. Ultimately, we argue that global climate justice requires that the intra-, inter- and transnational scales are all addressed.

Chapter 3, "Diverging geographies: Indigenous peoples, climate change and the UN COP summits," builds on the framework for global climate justice that combines moral and geographical scales by exploring how unequal impacts of climate change manifest *geographically* (Chatterton et al., 2013, p. 603) *in* frontline indigenous communities and *at* international climate summits. Empirically, testimonials shared by indigenous activists on the impacts of climate change are combined with our participant observations at COP17 and COP21. To begin, we consider research in a range of disciplines that frames climate change as a problem of scale and, in some instances, seeks to overcome this problem by bringing indigenous perspectives, particularly traditional ecological knowledge, into conversation with Western research paradigms. Next we delve deeper into the geographical dimensions of climate change by looking more carefully at three distinct scales: the geographies where indigenous peoples live; the spatial divisions inside UN summits; and the geography of protest in climate justice actions outside the summits. We conclude by discussing how these diverging spaces relate to each other in terms of power and opportunities to enhance indigenous access to media coverage and political arenas.

Chapter 4, "Summit journalism, indigenous peoples and digitalization: A media ecology perspective" takes a closer look at the diversity of the current media landscape and the place of indigenous activists within the media ecology that surrounds UN climate summits. We are particularly interested in parts of this media ecology where indigenous peoples are rendered *invisible*, made *visible* or strive to make themselves visible. Theoretically, the chapter builds on and develops media ecology theories (e.g. McLuhan, 1964; McLuhan & McLuhan, 1988; Postman, 2000; Scolari, 2013; Tracy, 2010, 2012). It also contributes to ongoing discussions of contemporary global power relations; particularly in terms of the potential for transnational groups to make their voices heard and take part in public discourse on par with everyone else. Building on our previous research, as well as interviews conducted in the field, this chapter illuminates the paradoxical situation indigenous activists find themselves in at UN COPs. On one hand, it is evident that they continue to view mainstream legacy media as crucial channels for disseminating their voices and experiences of climate change. On the other hand, they express collective frustration with the limited attention indigenous experiences of climate change receive in mainstream legacy media. We then ask how indigenous activists cope with this paradoxical situation? To find out, we consider two of the strategies indigenous activists employ to counter underrepresentation in legacy news media. The first is to turn to social media as an alternative source of pertinent news and information. The second strategy is media production, using digital tools to facilitate the self-representation of indigenous climate issues.

Chapter 5, "Activism, agonism, agency: Indigenous peoples, media witnessing, and the political game of the summits" builds on Chantal Mouffe's (2005, 2013) understanding of politics versus the political and elaborates on an agonistic (Mouffe, 2013) rather than a deliberative (see Habermas, 1994) view of democracy. Such a view is more in tune with justice perspectives, than consensus-oriented deliberative approaches. Empirically, the chapter builds on interviews and observations from two COPs (Durban, 2011 and Paris, 2015); quantitative data on newspaper material from COP coverage in 11 countries (2009, 2011, 2013, 2015); qualitative analysis of newspaper material from Sweden and Canada from COP15 in Copenhagen (2009) and COP21 in Paris (2015); and social media output from COP21 in Paris. We discuss media reporting on "politics" (the institutionalized side) and "the political" (the ideological side). The discussion spans how summit "politics" as a win-or-lose-game clash with ideological expressions ("the political") made by groups outside of institutional (inter-)national politics. We focus specifically on the ideological struggles that take place around the meetings, the radical politics of activists, and their views on rights and on media in relation to rights. The roles and views of indigenous activists are centred throughout. Interviewees state that they are being blinded and muffled in the political process, if/when they have access to it at all, and similar patterns can concurrently be found in media reporting. Zooming in on opportunities for capturing political aspects of climate change and indigenous peoples we target media witnessing—instances when indigenous peoples do have a voice in the media—and analyze how it plays out in diverse media contexts; particularly in mainstream media (edited media witnessing) and social media, such as YouTube and Facebook (un-/self-edited media witnessing). There are lessons to be learned from the mainstream as well as the social media material, which we discuss further in Chapter 6.

Chapter 6, "(Dis)connections: Particularism versus universalism, and transnational solidarity" concludes with a discussion of the various (dis)connections revealed in the previous chapters: (dis)connections regarding scales of justice; various forms of media, specifically indigenous peoples and mainstream media; indigenous peoples and official summit politics; indigenous peoples, rights and justice; and indigenous peoples and political discourses in the media. We subsequently zoom-in on how particularism and universalism can be understood based on the experiences and practices of indigenous peoples, specifically in relation to how media and politics focus on identity versus status. We then outline a practice of transnational solidarity that holds potential to bridge the disconnections we discovered. Empirically, the chapter focuses on input from interviewees on solidarity, on language in the final Paris Agreement (2015), and on diverse international declarations, conventions and covenants that pertain to human rights and indigenous peoples. We conclude by discussing how the media could implement a solidarian mode of communication in order to provide just reporting on climate change and indigenous peoples in a globalizing world.

Notes

1. These different aspects can be distinguished by differentiating between "natural" and "human-induced/anthropogenic" changes (*Ibid.*). An alternative term, "global warming" refers to human-induced climate changes caused by the emission of greenhouse gases (*Ibid.*). Another concept is the "greenhouse effect", which conveys a more limited meaning of the physical mechanics of planetary warming caused by the radiative heating of the earth's atmosphere by greenhouse gases (see Hulme, 2009, p. xxxix). In this book, we primarily use the term climate change. Global warming or greenhouse effect appear in instances where we are referring to precisely these phenomena, or in cases where the terms are used by the scholars or interviewees quoted.
2. For further discussion of media ecology theory, see Chapter 4 and Roosvall and Tegelberg, 2016.
3. The sample includes newspaper material from more than 20 countries, representing six continents, gathered by members of the research network. The participating countries have varied slightly over the years, see Eide, Kunelius and Kumpu (2010), Eide and Kunelius (2012), and Kunelius, Eide, Tegelberg and Yagodin (2017) for country lists and codebooks. In each country, two newspapers were monitored over a three-week period during the selected COPs (2007, 2009, 2011, 2013, 2015). Coding of the material was done using a joint codebook (see Eide, Kunelius and Kumpu, 2010; Eide and Kunelius, 2012). Both authors have been part of the MediaClimate network since it was established to study global media coverage of climate change in 2009. The network is led by Elisabeth Eide and Risto Kunelius. For more information on its activities and publications see: https://mediaclimate.net/publications.
4. The newspapers included in our main sample from these two countries are: *Aftonbladet* and *Dagens Nyheter* (Sweden) and *Globe and Mail* and *Toronto Star* (Canada).
5. In Tables 1.1 and 1.2 we focus on the degree of political influence/access and on the (political) roles the interviewees had at the summits. We also identify the (political) roles that brought them to the summit (based on the information they provided during the interviews). In other contexts, their roles may differ.
6. Näkkäläjärvi was on study leave from her position at Yle Sápmi at the time of the interview.

7. Most interviews were conducted in English. One was conducted in English and French (Diel Mochire) with the help of a French language translator. One was conducted in a combination of English and Spanish (Tito Puanchir), one was conducted in Swedish (Per Jonas Partapuoli) and one was conducted in a combination of Swedish and Norwegian (Áile Jávo).
8. "Pygmy" is sometimes viewed as a derogatory term. Here the interviewee self-identified as Pygmy.

References

Agyeman, J., Bullard, R., & Evans, B. (2003). *Just sustainabilities: Development in an Unequal World*. Cambridge, MA: MIT Press.

Agyeman, J., Cole, P., Haluza-DeLay, R., & O'Riley, P. (Eds.). (2009). *Speaking for ourselves: Environmental justice in Canada*. British Columbia: University of British Columbia Press.

Alia, V. (2010). *The new media nation: Indigenous peoples and global communication*. London: Berghan Books.

Altheide, D. H. (1997). The news media, the problem frame, and the production of fear. *The Sociological Quarterly*, *38*(4), 647–668.

Anderson, M., & Robertson, C. (2011). *Seeing red: A history of natives in Canadian newspapers*. Manitoba: University of Manitoba Press.

Appadurai, A. (1996). *Modernity at large: Cultural dimensions of globalization*. Minneapolis, MN: University of Minnesota Press.

Askanius, T., & Uldham, J. (2011). Online social media for radical politics: Climate change activism on YouTube. *International Journal of Electronic Governance*, *4*(1), 69–84.

Bennett, W. L., & Segerberg, A. (2012). The logic of connective action: Digital media and the personalization of contentious politics. *Information, Communication and Society*, *15*(5), 739–768.

Bennett, W. L., & Segerberg, A. (2013). *The logic of connective action: Digital media and the personalization of contentious politics*. New York, NY: Cambridge University Press.

Boyce, T., & Lewis, J. (Eds.). (2009). *Climate change and the media*. New York, NY: Peter Lang.

Boykoff, M. T. (2011). *Who speaks for the climate: Making sense of media reporting on climate change*. Cambridge: Cambridge University Press.

Boykoff, M. T., & Boykoff, J. M. (2004). Balance as bias: Global warming and the US prestige press. *Global Environmental Change*, *14*(2), 125–136.

Brülde, B., & Duus-Otterström, G. (2015). *Klimatetik: Rättvisa, politik och individens ansvar* [Climate ethics: Justice, politics, and the responsibility of the individual]. Stockholm: Thales.

Bullard, R. (1990). *Dumping in Dixie: Race, class, and environmental quality*. San Francisco, CA: Westview Press.

Burke, M. B., Miguel, E., Satyanath, S., Dykena, J. A., & Lobell, D. B. (2009). Warming increases the risk of civil war in Africa. *Proceedings of the National Academy of Sciences*, *106*(49), 20670–20674.

Bryman, A. (1996). *Social research methods*. London: Oxford University Press.

Carbaugh, D., & Cerulli, T. (2013). Cultural discourses of dwelling: Investigating environmental communication as a place-based practice. *Environmental Communication: A Journal of Nature and Culture*, *7*(1), 4–23.

Carvalho, A. (2005). Representing the politics of the greenhouse effect: Discursive strategies in the British media. *Critical Discourse Studies*, *2*(1), 1–29.

Carvalho, A. (2007). Communicating global responsibility? Discourses on climate change and citizenship. *International Journal of Media and Cultural Studies*, *3*(2), 180–183.

Chatterton, P., Featherstone, D., & Routledge, P. (2013). Articulating climate justice in Copenhagen: Antagonism, the commons, and solidarity. *Antipode*, *45*(3), 602–620.

Chouliaraki, L. (2008). The media as moral education: Mediation and action. *Media, Culture and Society*, *30*(6), 831–852.

Christensen, C. (2011). Twitter revolutions? Addressing social media and dissent. *Communication Review*, 14(3), 155–157.

Christensen, M. (2013). Arctic climate change and the media: The news story that was. In M. Christensen, A. E. Nilsson, & N. Wormbs (Eds.), *Media and the politics of Arctic climate change: When the ice breaks* (pp. 26–51). Basingstoke: Palgrave.

Corbett, J. B. (2012). *Communicating nature: How we create and understand environmental messages*. Washington, DC: Island Press.

Cottle, S. (2009). Global crises in the news: Staging new wars, disasters, and climate change. *International Journal of Communication*, 3, 494–516.

Cottle, S., & Lester, L. (2011). *Transnational protests and the media*. New York, NY: Peter Lang.

Couldry, N. (2012). *Media, society, world: Social theory and digital media practice*. Cambridge: Polity.

Davidsen, C. (2011). Amazon struggles in the global media age: Framing and discourses in environmental conflict. In S. Cottle & L. Lester (Eds.), *Transnational protests and the media* (pp. 172–184). New York, NY: Peter Lang.

Dayan, D., & Katz, E. (1992). *Media events: The live broadcasting of history*. Cambridge: Harvard University Press.

de Onís, K. (2012). Looking both ways: Metaphor and the rhetorical alignment of intersectional climate justice and reproductive justice concerns. *Environmental Communication*, 6(3), 308–327.

Doyle, J. (2011). *Mediating climate change*. London: Ashgate.

Drugge, A. (Ed.). (2016). *Ethics in indigenous research: Past experiences, future challenges*. Umeå: Vaaartoe—Centre for Sami Research.

Eide, E., Kunelius, K., & Kumpu, V. (Eds.). (2010). *Global climate, local journalisms: A transnational study of how media make sense of climate summits*. Bochum: Projekt Verlag.

Eide, E., & Kunelius, R. (Eds.). (2012). *Media meets climate: The global challenge for journalism*. Sweden: Nordicom.

Eide, E., Kunelius, R., & Philips, A. (Eds.). (2008). *Transnational media events: The Mohammed cartoons and the imagined clash of civilizations*. Sweden: Nordicom.

Endres, D. (2009). From wasteland to waste site: The role of discourse in nuclear power's environmental injustices. *Local Environment: The International Journal of Justice and Sustainability*, *14*(10), 917–937.

Entman, R. (1993). Framing: Towards a clarification for a fractured paradigm. *Journal of Communication*, *43*(4), 51–58.

Entman, R. (2007). Framing bias: Media in the distribution of power. *Journal of Communication*, *57*(2), 163–173.

EWContributor. (2015, November 29). 10,000 form human chain in Paris demanding world leaders keep fossil fuels in the ground. *EcoWatch*. Retrieved from http://www.ecowatch.com/10-000-form-human-chain-in-paris-demanding-world-leaders-keep-fossil-f-1882127390.html

Fontaine, T. (2015, November 30). Indigenous leaders, activists in Paris for climate change talks. *CBC News—Indigenous*. Retrieved from http://www.cbc.ca/news/aboriginal/indigenous-leaders-activist-paris-climate-talks-1.3343927

Fraser, N. (2000). Rethinking recognition: Overcoming displacement and reification in cultural politics. *New Left Review*, *3*, 107–120.

Fraser, N. (2008). *Scales of justice: Reimagining political space in a globalizing world*. New York, NY: Columbia University Press.

Fraser, N. (2014). Transnationalizing the public sphere: On the legitimacy and efficiency of public opinion in a post-Westphalian world. In K. Nash (Ed.), *Transnationalizing the public sphere* (pp. 8–42). Cambridge: Polity.

Frosh, P., & Pinchevski, A. (Eds.). (2009). *Media witnessing: Testimony in the age of mass communication*. Basingstoke: Palgrave.

Gavin, N., & Marshall, T. (2011). Climate change and international protest at Copenhagen: Reflections on British television and the web. In S. Cottle, & L. Lester (Eds.), *Transnational protests and the media* (pp. 197–212). New York, NY: Peter Lang.

Georgiou, M. (2011). Diaspora, mediated communication and space. In M. Christensen, A. Jansson, & D. A. Gozdecka (Eds.), *Online territories* (pp. 205–221). New York, NY: Peter Lang.

Giddens, A. (2009). *The politics of climate change*. Cambridge: Polity.

Habermas, J. (1994). Three normative models of democracy. *Constellations*, *1*(1), 1–10.

Harvey, D. (1996). *Justice, nature and the geography of difference*. Oxford: Blackwell.

Hulme, M. (2009). *Why we disagree about climate change: Understanding controversy, inaction and opportunity*. Cambridge: Cambridge University Press.

Indigenous Environmental Network. (2015, November 30). In *Facebook* [Group page]. Retrieved from https://www.facebook.com/ienearth/videos/10153285239180642/

Iorio, S. H. (2011). Qualitative method journalism. In S. H. Iorio (Ed.), *Qualitative research in journalism: Taking it to the streets* (pp. 3–20). New York, NY: Routledge.

Jackson, G., & Parry, M. (2015, November 29). Paris human chain gives emotional jolt to climate summit. *Agence France-Presse*. Retrieved from https://www.yahoo.com/news/world-wide-rallies-demand-action-climate-change-043006042.html?ref=gs

Johnson, S. (2009). Climate change and global justice: Crafting fair solutions for nations and peoples. *Harvard Environmental Law Review*, *33*(2), 297–301.

Kals, E., & Russell, Y. (2001). Individual conceptions of justice and their potential for explaining proenvironmental decision making. *Social Justice Research*, *14*(4), 367–385.

Klein, N. (2014). *This changes everything: Capitalism vs. the climate*. New York, NY: Simon & Schuster.
Krzyżanowski, M. & Forchtner, B. (2016). Theories and concepts in critical discourse studies: Facing challenges, moving beyond foundations. *Discourse & Society*, 27(3), 253–261.
Laclau, E., & Mouffe, C. (1985). *Hegemony and socialist strategy: Towards a radical democratic politics*. New York, NY: Verso.
LaDuke, W. (1994). Traditional ecological knowledge and environmental futures. *Colorado International Environmental Law and Policy*, 5, 127–148.
LaDuke, W. (1999). *All our relations: Native struggles for land and life*. Cambridge, MA: South End Press.
Ledman, A. (2015). Turning the gaze from "them" to "us"—Challenging the mainstream in indigenous research. In P. Sköld, M. Sandström, & M. Bolaane (Eds.), *Under the same sun: Parallel issues and mutual challenges for San and Sami Peoples and research* (pp. 159–168). Umeå: Vaartoe—Centre for Sami Research.
Logan, R. K. (2007). The biological foundation of media ecology. *Explorations in Media Ecology*, 6(1), 19–34.
Lövbrand, E., Beck, S., Chilvers, J., Forsyth, T., Hedre, J., Hulme, M., Lidskog, R., Vasileiadou, E. (2015). Who speaks for the future of earth? How critical social science can extend the conversation on the Anthropocene. *Global Environmental Change*, 32, 211–218.
McLuhan, M. (1964). *Understanding media: The extensions of man*. New York, NY: McGraw-Hill.
McLuhan, M., & McLuhan, E. (1988). *The laws of media: The new science*. Toronto: University of Toronto Press.
Mouffe, C. (2005). *On the political*. London: Routledge.
Mouffe, C. (2013). *Agonistics: Thinking the world politically*. London: Verso.
Olausson, U. (2014). The diversified nature of "domesticated" news discourse: The case of climate change in national news media. *Journalism Studies*, 15(6), 711–725.
Page, E. A. (2006). *Climate change, justice and future generations*. Cheltenham, UK: Edward Elgar Publishing.
Painter, J., Erviti, M. C., Fletcher, R., Howarth, C., Kristiansen, S., Leon, B., Ouakrat, Russell, A., Schäfer, M. (2017). *Something old, something new: Digital media and the coverage of climate change*. Oxford: Reuters Institute for the Study of Journalism.
Peters, J. D. (2001). Witnessing. *Media, Culture and Society*, 23(6), 707–723.
Pezzullo, P., & Sandler, R. (2007). *Environmental justice and environmentalism: The social justice challenge to the environmental movement*. Cambridge, MA: MIT Press.
Plaut, S. (2017). Reshaping the borders of journalism: Lessons learned from transnational peoples' journalism. *Journalism Practice*, 11(1), 1–32.
Postman, N. (2000). The humanism of media ecology. *Proceedings of the Media Ecology Association*, 1(1), 10–16.
Prince, G., & Rayner, S. (2007, October). *The wrong trousers: Radically rethinking climate policy*. Joint discussion paper of the James Martin Institute for Science and Civilisation, Oxford, and the MacKinder Centre for the Study of Laong-Wave Events, London School of Economics.

Rantanen, T. (2005). *The media and globalization*. London: Sage.

Rao, S. (2015). Practices of Indian journalism: Justice, ethics and globalization. In S. Rao & H. Wasserman (Eds.), *Media ethics and justice in the age of globalization* (pp. 121–138). London: Palgrave Macmillan.

Reese, G. (2016). Common human identity and the path to global climate justice. *Climatic Change, 134*(4), 521–531.

Reese, S. D., & Danielian, L. H. (1989). Intermedia influence and the drug issue. In P. J. Shoemaker (Ed.), *Communication campaigns about drugs* (pp. 29–45). Hillsdale, NJ: Elbaum.

Reser, J. P., Morissey, S. A., & Ellul, M. (2011). The threat of climate change: Psychological response, adaptation and impacts. In I. Weissbecker (Ed.), *Climate change and human well-being* (pp. 19–42). New York, NY: Springer.

Robertson, A. (2015). Global news: Reporting conflicts and cosmopolitanism. New York, NY: Peter Lang.

Roosvall, A. (2010). Image-Nation: The national, the cultural and the global in foreign news slideshows. In A. Roosvall & I. Salovaara-Moring (Eds.), *Communicating the nation: National topographies of global media landscapes* (pp. 215–236). Gothenburg: Nordicom.

Roosvall, A. (2016). Journalism, climate change, justice and solidarity: Exploring attitudes in editorials surrounding the IPCC reports. In E. Eide, R. Kunelius, M. Tegelberg, & D. Yagodin (Eds.), *Media and global climate knowledge: Climate journalism and the IPCC* (pp. 129–150). New York: Palgrave.

Roosvall, A., & Tegelberg, M. (2012). Misframing the messenger: Scales of justice, traditional ecological knowledge and media coverage of indigenous peoples and climate change. In E. Eide & R. Kunelius (Eds.), *Media meets climate: The global challenge for journalism* (pp. 297–312). Gothenburg: Nordicom.

Roosvall, A., & Tegelberg, M. (2013). Framing climate change and indigenous peoples: intermediaries of urgency, spirituality and de-nationalization. *International Communication Gazette, 75*(4), 392–409.

Roosvall, A., & Tegelberg, M. (2015). Media and the geographies of climate justice: Indigenous peoples, nature and the geopolitics of climate change. *TripleC: Communication, Capitalism & Critique, 13*(1), 39–54.

Roosvall, A., & Tegelberg, M. (2016). Natural ecology meets media ecology: Indigenous activists' views on nature and media. In H. Graf (Ed.), *Environment in the age of the internet: Activists, communication, and the digital landscape* (pp. 75–104). Cambridge: Open Book Publishers.

Sampei, Y., & Aoyagi-Usui, M. (2009). Mass-media coverage, its influences on public awareness of climate-change issues, and implications for Japan's national campaign to reduce greenhouse gas emissions. *Global Environmental Change*, 19(2), 203–212.

Schmidt, A., & Schäfer, M. (2015). Constructions of climate justice in German, Indian and US media. *Climatic Change, 133*(3), 535–549.

Schubert, R., & Schnellhuber, H. J. (2009). *Climate change as a security risk*. London: Routledge.

Scolari, C. A. (2013). Media evolution: Emergence, dominance, survival and extinction in the media ecology. *International Journal of Communication, 7*(1), 1418–1441.

Segerberg, A., & Bennett, L. (2011). Social media and the organization of collective action: Using Twitter to explore the ecologies of two climate change protests. *Communication Review, 14*, 197–215.

Sen, A. (2009). *The idea of justice*. Cambridge: Harvard University Press.

Shepard, P., & Corbin-Mark, C. (2009). Climate justice. *Environmental Justice, 2*(4), 163–166.

Shue, H. (2014). *Climate justice: Vulnerability and protection*. Oxford: Oxford University Press.

Sikor, T., Martin, A., Fisher, J., & He, J. (2014). Toward an empirical analysis of justice in ecosystem governance. *Conservation Letters: A Journal of the Society for Conservation Biology, 7*(6), 524–532.

Smith, L. T. (1999). *Decolonizing methodologies: Research and indigenous peoples*. London: Zed Books.

Sowards, S. K. (2012). Environmental justice in international contexts: Understanding intersections for social justice in the twenty-first century. *Environmental Communication: A Journal of Nature and Culture*, 6(3), 285–289.

Swim, J., Clayton, S., Doherty, T., Gifford, R., Howard, G., Reser, J., … Weber, E. (2009). *Psychology and global climate change: Addressing a multifaceted phenomenon and set of challenges*. Washington, DC: American Psychological Association.

Szasz, A. (1994). *Ecopopulism: Toxic waste and the movement for environmental justice*. Minneapolis, MN: University of Minnesota Press.

Tauli-Corpuz, V. (2009). Introduction. In V. Tauli-Corpuz & R. de Chavez (Eds.), *Guide on climate change and indigenous peoples* (2nd ed.). Baguio City: Tebtebba Foundation.

Thorsen, K., Edgerly, S., Kliger-Vilenchik, N., Xu, Y., & Wang, L. (2016). Seeking visibility in a big tent: Digital communication and the people's climate March. *International Journal of Communication, 10*, 4784–4806.

Tracy, C. (2010). A quantum exploration of the news ecosystem. *Explorations in Media Ecology*, 9(3), 193–198.

Tracy, C. (2012). Ecological journalism and the role of perception. *Explorations in Media Ecology, 11*(2), 131–142.

Uldham, J., & Askanius, T. (2013). Online civic cultures: Debating climate change activism on YouTube. *International Journal of Communication, 7*, 1185–1204.

United Nations Permanent Forum on Indigenous Issues. (n.d.). *Who are indigenous peoples?* Retrieved from http://www.un.org/esa/socdev/unpfii/documents/5session_factsheet1.pdf

Urry, J. (2011). *Climate change and society*. Cambridge: Polity Press.

Westra, L. (2007). *Environmental justice and the rights of indigenous peoples: International and domestic law perspectives*. London: Earthscan Publishers.

Williams, H. T. P., McMurray, J. R., Kurtz, T., & Lambert, H. F. (2015). Network analysis reveals open forums and echo chambers in social media discussions of climate change. *Global Environmental Change, 32*(2015), 126–138.

Winter-Jørgensen, M., & Phillips, L. (2002). *Discourse analysis as theory and method*. London: Sage.

World Resources Institute. (2013). Declaration on climate justice. *Mary Robinson Foundation Climate Justice and the World Resources Institute*. Retrieved from http://www.wri.org/sites/default/files/declaration_on_climate_justice_0_0.pdf

Ytterstad, A., & Russell, A. (2012). Pessimism of the intellect and optimism of the will: A Gramscian analysis of climate justice in summit coverage. In E. Eide & R. Kunelius (Eds.), *Media meets climate: The global challenge for journalism* (pp. 247–262). Gothenburg: Nordicom.

· 2 ·

WHAT IS CLIMATE JUSTICE?

Justice, Climate and the Media

> I want to invite climate justice into Chukotka region … to help us to protect the territory and save our lands.
>
> —Olga Atsynga Letykai Csonka, Chukchi people (Russia, COP21)

When Olga Atsynga Letykai Csonka talks about inviting climate justice into the Chukotka region in north-eastern Russia, she alludes to a dearth of ethics, global equality, human rights, democratic accountability, issues of participation and historical responsibility when it comes to climate change effects and how they need to be tackled there (see Chatterton, Featherstone, & Routledge, 2013; de Onís, 2012; Page, 2006; Shepard & Corbin-Mark, 2009; Shue, 2014). It is a *locally* rooted experience of vulnerability, injustice and lack of responsibility from more privileged groups. At the same time it is, as we shall see, a *common* experience in a worldwide, *transnational* community of diverse indigenous groups, trying to protect their territories and save their lands on all continents.

In order to understand *climate* justice we need to delineate how justice can be understood more broadly. Thus we begin this largely theoretical chapter by discussing *the substance of justice* and connected ways to *amend injustice* (Fraser, 2008). Subsequently we detail how *responsibility* for justice can be understood.

Who is responsible for amending injustice? And how does responsibility relate to the media? These issues are thereafter related to *the framing of justice*, focused on geographical justice scales, evoked by a need to rethink justice in relation to globalization. A crucial instance of injustice in a globalizing age is misframing (Fraser, 2008), which we discuss throughout the book in relation to the media as framing agents.

The bulk of the literature on climate justice understands the concept as having to do with justice between nation-states on an inter-national scale (e.g. Michaelova & Michaelova, 2015; Page, 2012; Schmidt & Schäfer, 2015; Shue, 2014). We add to this by underlining intranational and transnational dimensions of climate justice, which are particularly relevant for indigenous peoples who are often transnational (e.g. Sami, Inuit, Tuareg) and simultaneously marginalized in the territories of the respective nation-states where they dwell (Roosvall, 2016; Roosvall & Tegelberg, 2015). These dimensions are concurrently crucial for all peoples in a world where connections are increasingly globalizing; specifically concerning a phenomenon such as climate change, which knows no borders, but has more severe effects in some areas than in others.

When detailing climate justice, we start with the *international* scale, where we discuss three principles generally used when negotiating the dissemination of blame and responsibility for causing and tackling climate change effects: The Polluter Pays Principle (PPP), the Beneficiary Pays Principle (BPP) and the Ability to Pay Principle (APP) (Brülde & Duus-Otterström, 2015). We subsequently move to the *intranational* scale, relevant to indigenous groups and their status within diverse nation-states, such as Letykai Csonka's Chukchi people in Russia, the Kichwa people of Sarayaku in Ecuador, the Māori in New Zealand, Cree in Canada and the United States, and the Kel Tamasheq people (Tuareg) in Algeria, Burkina Faso, Libya, Mali and Niger. Then we zoom-in on *transnational and global scales*, which are often strongly connected to intranational scales in indigenous struggles for justice. We discuss the necessity and possibilities of transnational climate justice, through the construction of a Common Human Identity (CHI) (Reese, 2016). Furthermore, we relate this to diverse intranational experiences and identity claims, and the possible tensions between geographical scales that this may bring. In conclusion, we argue that global climate justice requires that the intra-, inter- and transnational levels are all addressed.

In addition to space, time plays a crucial role for climate justice, often conceptualized as intergenerational climate justice (Page, 2006), or intergenerational *responsibility* in relation to climate change (Hayden, 2010), with varying emphasis on past, present, and future. We integrate views on time into each of the subsections covering distinct geographical scales.

As we navigate towards a deeper understanding of climate justice we draw not only on previous studies and theoretical trajectories, but also on knowledge imparted by the interviewed indigenous activists, including their understandings of climate justice and experiences of climate injustice. Accounts from the interviewees are placed in dialogue with theories on justice and responsibility. Thus, we listen to the voices of indigenous activists as subjects of climate justice (both in terms of their subjection to injustice and their actions as subjects for justice). We view them, moreover, as subjects that are defined by and simultaneously define climate justice discourse. Their accounts help pinpoint the ruptures in previous understandings of climate justice and contribute to outlining what still needs to be explored and included in such understandings, as well as in understandings of justice at large, and how this relates to the media. The interviewees contribute in this way from their often concurrently intranational and transnational points-of-view, from their lived through and passed-down perspectives on nature, as well as from their experiences of colonization, of suppression and of having to fight for the very right to have rights (Arendt, 1998; see also Chapter 5). They thereby inform our understanding of relationships between climate change, people and rights, how this can and should be reconceptualised, and more particularly how climate justice needs to be rethought to encompass situated local experiences *and* the transnational scale.

Olga Atsynga Letykai Csonka talks about how climate justice is needed to help her people protect their territories and save their lands. With this as a starting point, in this chapter and throughout this book we discuss how, if applied in time, indigenous perspectives on climate justice—and activism directed towards it—could help save *all* lands, not least their own. Thus, this chapter addresses the questions: How can we in the most just way understand and approach issues of climate change in globalizing and unequal times? How can we most constructively understand and approach *responsibility* for justice regarding the transnational phenomenon of climate change, and what is the media's role in relation to this? How do issues concerning intra- and transnational groups, and more particularly indigenous peoples, relate to a climate justice literature thus far dominated by the international geographical scale? And finally, how can these dimensions be better integrated and what could be gained from such an integration in terms of justice?

The Justice in Climate Justice

The case of indigenous peoples, climate justice and the media is particularly well suited for elaborating on theories of justice in a globalizing world. Understandings of justice and globalization on a more general level, as detailed in this section, are deepened when they are related to the transnational phenomenon of climate change (see also Tokar, 2014); to intra- and transnational indigenous peoples; to media which are often still connected to the national level—through language, ownership and audiences—but also operate along other geographical scales (local, transnational); and finally to how all of these aspects intersect. Departing from Rawls' (1971, p. 7) nation-state based theories, we thus heed the re-theorizing of justice in a globalizing world in political philosophy (Fraser, 2008, 2014; Nash, 2014; Sen, 2009; Young, 2013) and in media studies (Couldry, 2012; Raboy & Mawani, 2014; Rao & Wassermann, 2015; Silverstone, 2007).

Normative justice addresses the obligation of humankind to humankind; it can be defined as "setting human affairs right in terms of human rights, human goods (humanity as the end), and human bonds" (Chen, 2008, p. 66). It is often focused on social institutions (Chen, 2008; Rawls, 1971, p. 6) and elaborates on *ideals* for justice. Rather than elaborating on normative ideals, *sociological theories on justice* focus on the importance of specific contexts in space and time and on how justice is viewed by different societies or stakeholders. Schmidt and Schäfer (2015, p. 536) recognize sociological perspectives on justice as acknowledging the "many objective ends, ultimate values, some incompatible with others, [which are] pursued by different societies at various times, or by different groups in the same society" (cf. Berlin, 2013). Such perspectives focus on the justifications and criticisms put forward by disputing actors and reconstruct shared values and principles (Boltanski & Thévenot, 2006; cf. Schmidt & Schäfer, 2015, p. 36). We argue it is imperative to combine normative and sociological approaches to justice in order to stipulate ideals *and* assess the actual implementation of in/justice in diverse contexts (Sen, 2009; cf. Rao, 2015, p. 122): e.g. time, space, indigenous/non-indigenous groups and diverse media. While the theories we draw on are mainly normative, we put them in dialogue with assessments from the indigenous activists we interviewed. These perspectives shed new light on the validity of these theories and thus contribute to their journey beyond the narrow Western and nation-state centred approaches.

Another crucial distinction among justice theories is between those focused on individual versus social justice (LeBar & Slote, 2016). Individual centred theories are often conceptualized as "moral", and focus on individual responsibility for employing justice and amending injustice (e.g. Silverstone, 2007). Social centred theories in turn (e.g. Rawls, 1971) focus more on collectives and institutions. In this book we consider both individual and social responsibility for justice, focusing to a large extent on democratic justice, which is concerned with participation in political processes (e.g. Fraser, 2008, p. 28). While we emphasize the collective level, we also acknowledge the importance of individual action as parts of collectives, and consider especially the actions of those individuals and collectives that inhabit powerful positions in the media (Silverstone, 2007, p. 153) and in politics.

The Substance of Justice

Fraser's (2008, p. 16) fundamental definition of justice is *parity of participation*: "justice requires social arrangements that permit all to participate as peers in social life. Overcoming injustice means dismantling institutionalized obstacles that prevent some people from participating on a par with others, as full partners in social interaction". There are three main types of such obstacles: (i) economic structures that hinder participation, i.e. distributive injustice or maldistribution; (ii) institutionalized hierarchies of cultural value that hinder participation, i.e. status inequality or misrecognition. These diverse forms of inequality are sometimes connected, but one is not merely, nor automatically, the result of the other. In the context of globalization, Fraser (2008, p. 17) adds a third structure that hinders participation on a par with others: (iii) political structures that prevent just political representation for certain groups, i.e. misrepresentation.

Following Sen's (2009, cf. Rao & Wassermann, 2015, p. 11) contention that there is a need to think deeply about *injustice* and not just postulating norms for justice, we distinguish between Fraser's three main remedies for injustice connecting them to the three aforementioned obstacles: (i) *redistribution* of means (economic remedy), (ii) *recognition* of identity/status (cultural remedy) and (iii) *representation* (political remedy) (Fraser, 2000, 2008, 2014; Fraser & Honneth, 2003; Fraser & Olson, 2008; Nash, 2014). Redistribution was employed in the post-WWII era, for instance, in states newly freed from colonization (Fraser, 2000). It subsequently took the backseat to recognition approaches, which came to dominate the identity politics of the 1990s.

Initially, recognition appeared in the form that Fraser (2000, 2008) calls *the identity model*, focusing, for example, on the right of ethnic groups to use their own language in schools and to express their culture in various ways.

Fraser (2000) later added that the recognition of status was as important as the recognition of identity. *The status model* means that subordination is tackled by "establishing the misrecognized party as a full member of society, capable of participating on a par with the rest" (Fraser, 2000, p. 113). Hence, it aims to "de-institutionalize patterns of cultural value that impede parity of participation and to replace them with patterns that foster it" (Fraser, 2000, pp. 114–115). This, in fact, comes very close to Fraser's most basic definition of justice overall; that is "parity of participation." Moreover, it coincides with calls for combinations of redistribution and recognition approaches, tackling economic and cultural injustices that are often interconnected. It is also noteworthy that in highly mediatized societies patterns that foster or impede parity are, to a large extent, expressed in the media. Remedy in the form of *representation*, in turn, was added by Fraser (2008, 2014) to cover participation on the official level. Representation means political representation. While these remedies for injustice complement each other, when strictly speaking about climate justice and what it means, our respondents evoke the redistribution aspect most often:

> … part of the general international legislation, or international law … could be used [to] reduce the climate change, reduce the temperature, and just give *the financial support* from the … developed states to the developing states—to the states in the South. (Dmitry Berezhkov, Itelmen people, Russia, COP21, emphasis added)

> Climate justice means having an equal treatment or *distribution of the resources that we have*. (Justeene Dammay, Tuwali people, Philippines, COP21, emphasis added)

> We have a right to live on a planet that is habitable and that can sustain us, the communities in the Arctic. I mean, I'm from a community in the Arctic. For instance, the young man that I'm travelling with, Esau, he's from a village called Shishmaref that is literally sinking into the ocean … where is his justice? Why does his community have to find a solution *to move their village for $250 million* and figure it out on their own, when the village is sinking because of the rising ocean levels, and it is not their [fault], and it is manmade? (Allison Akootchook Warden, Iñupiaq people, USA, COP21, emphasis added)

In Chapter 5 we consider input from interviewees regarding recognition and representation while discussing activism and political processes. Until then, here is one example of each.

Recognition:

> Climate justice, I think it needs to have a strong link to those communities that are on the front lines of climate change, and ... the differential impacts upon those people ... marginalized peoples, indigenous peoples ... *recognizing* that it's through colonization ... at least in the case of the Pacific ... You know most of the damage done to the environment is [done] by the developed countries. (Teanau Tuiono, Māori, New Zealand and Cook Islands, COP21, emphasis added)

Representation (political participation):

> We are just an observer; we cannot *participate* in the process ... we can only lobby and submit our statements, or position papers to different delegates of the countries. (Anthony Debbarma, Borok people, India, COP21, emphasis added)

Aside from the redistribution-recognition-representation triad, yet another triad of concepts helps define justice: "the what", "the who" and "the how" of justice (Fraser, 2008, pp. 53–54). The "what" of justice concerns whether the justice issue is defined by maldistribution, misrecognition or misrepresentation (political domination). The "who" concerns who counts as the subject of justice in a given situation: "Whose interests and needs deserve consideration? Who belongs to the circle of those entitled to equal concern?" (Fraser, 2008, p. 54). "How" in turn concerns the ways disputes regarding the "what" and the "who" should be resolved. The "what" has been theorized and exemplified above. The "who" will be theorized in the sections below on responsibility, respectively the framing of justice. In Chapters 3 and 5, we elaborate on the "how" in relation to the political geography of the summits, to activism and to the potential for indigenous peoples to take part in political decision-making. A focus on the "how" of justice connects to issues of procedural justice (Silverstone, 2007, p. 149) and implies an emphasis on democratic justice (Fraser, 2008, pp. 7, 28–29) rather than "merely" social justice, on the collective level.

Responsibility

In a world marked by competition over economic, natural and information resources, rights may not always be accessible in reality, nor justice implemented only because it is just. Rather for rights to be implemented, responsibility is needed (Roosvall, 2016). Ayish (2015, p. 139), in fact, understands justice *as* responsibility. Responsibility is also something many of the activists we interviewed refer to spontaneously when discussing climate justice:

Interviewer:	How do you define climate justice?
India Logan Riley:	Fairness and equity I guess, and also those who have … historical *responsibility* for where we are now, to step up to the plate and act in accordance with acknowledging that *responsibility* and compensating for it in some way … Fairness, equality, and *responsibility*. (India Logan Riley, Māori, New Zealand, COP21, emphasis added)

India Logan Riley calls for action in relation to responsibility. This is something that Young (2013, p. 144) elaborates on when she expresses what can be expected of us regarding issues of injustice: "we can and should be criticized for not taking action, not taking enough action, taking ineffective action, or taking action that is counterproductive". She adds that two groups are particularly responsible for taking issue with injustice: persons privileged by the structures, and victims of injustice (Young, 2013, p. 145). The latter are considered responsible for taking action because they are directly affected. Since power and ability are also recognized as important for determining who are most responsible for action (Young, 2013, p. 147; see also Silverstone, 2007, p. 153, on substantive responsibility), as implied by Young's pinpointing of those privileged by the structures, we argue that news media—as powerful framing agents—must shoulder extensive responsibility. At the same time, placing responsibility on victims, as Young does, becomes problematic in today's media ecologies.[1] The power to take issue today depends largely on access to media, which are channels for discursive action. It is via these channels that issues of injustice are largely covered, and consequently it is there, to a significant extent, that required remedies could—at least potentially—be outlined. Media ecologies are however harsh for indigenous "victims" aiming to spread awareness of climate change effects in their communities (Roosvall & Tegelberg, 2016). While indigenous peoples eagerly seek the mainstream media's attention, these media largely ignore their voices and perspectives, particularly regarding redistribution and political representation. Recognition of status has also been rare (Roosvall & Tegelberg, 2015, 2016).

Accordingly, while indigenous activists strive to have a voice regarding the climate changes experienced and the political changes needed, they expect others to take responsibility for this unjust situation:

> Climate justice means to me that those affected by climate in the frontline communities are not asked to pay for, you know, and take *responsibility* for these actions that we had no part of, so that we have support for communities that are dealing with these affects, from these multinational corporations and these other things that are, you

> know, warming the planet, so that they take *responsibility* for what they are doing and not the communities having to scramble and try to figure out what to do. (Allison Akootchook Warden, Iñupiaq people, USA, COP21, emphasis added)

The question of who is responsible for taking action on injustice, discussed by Akootchook Warden above, can be discussed in terms of *the all-affected* versus *the all-subjected principles* (Fraser, 2008; Owen, 2014). The all-affected principle includes only those directly affected by unfair situations: flooding, melting ice, drought, among others, whereas the all-subjected principle encompasses everyone implicated in the structures to which the injustice is connected. As Fraser (2008, p. 96) puts it: "what turns a collection of people into fellow members of a public is not their shared citizenship, or co-imbrication in a causal matrix, but rather their joint subjection to a structure of governance that sets the ground rules for their interaction". The all-affected principle has been criticized for ultimately risking to evoke the "butterfly effect", where all actions may in the end affect all of the future, making it impossible to determine the who of justice (Owen, 2014).

The all-subjected principle is more precise in that it pinpoints governance and includes *the construction of frames of participation*, indicating that everyone potentially subjected must be able to *make decisions on how decisions are to be made* (i.e. who can take part in the decision-making process). Concerning climate change, many people thus become part of the justice issue in a less vague way, since most people can be considered subjected under the UN COP umbrella. Hence, everyone should in principle be able to affect who can participate in decision-making.[2] We add that the media is likewise subjected to the structures of governance in its own way, while simultaneously reporting on it. Silverstone (2007, p. 10) underlines the role of media as a moral force in cases where "the world in its otherness is at its most visible", which is certainly relevant in the (few) occasions when indigenous peoples do get media attention in relation to climate change.[3] Silverstone (2007, p. 11) continues by noting that technologies disconnect as well as connect. That is, they often undermine the expectation of representation and *responsibility* that action in face-to-face settings would demand. Hence, how media represents the other is a key issue for any project striving for a more virtuous ethical public space. This is pertinent to the situation concerning indigenous peoples, climate change and the media ecologies that are crucial for shaping public knowledge of these issues.

Young (2013: 124) draws on what she calls the social connection model for responsibility when explaining how responsibility – like justice – needs to extend across borders. Responsibility for justice cannot be restricted to members of a nation-state, but is rather determined by the social connections between people (such as through the all-subjected principle). In this transnational context, a responsible person "tries to deliberate about options before acting, makes choices that seem to be the best for all affected, and worries about how the consequences of his or her actions may adversely affect others" (Young, 2013, p. 25). Assigning responsibility to someone is to say they have a job to do; they have to act (Nussbaum, 2013; Young, 2013). *Political* responsibility for instance entails monitoring societal institutions and their effects "to make sure they are not grossly harmful, and maintaining organized public space where such watching and monitoring can occur and citizens can speak publicly and support one another in their efforts to prevent suffering" (Nussbaum, 2013, p. xv). One of the media's main roles in society is such monitoring. Concurrently, the media itself is a powerful social institution (Silverstone, 2007, p. 144). Thus the media must both monitor and be monitored, making its role in these processes particularly significant and in need of further scrutinizing.

According to the *substantive aspect of responsibility*, media and politicians are among those that can be held more responsible than others. They are made accountable by virtue of their existence on this planet, and their existence as agents *proportionate to their power* (Silverstone, 2007, p. 153). This entails responsibility for the existence of the other, as that of a parent for a child or a public figure for their constituents. It should be accompanied by the *formal aspect of responsibility*, which entails responsibilities for one's own actions and direct accountabilities (Silverstone, 2007, p. 152).[4] Áile Jávo, President of the Sami Council, connects to substantive responsibility when she assigns the media responsibility for holding leaders accountable:

> If one could make a wish: that the media would hold the leaders responsible, and actively asked why … You see … climate change affects indigenous peoples, why don't you do anything about that? (Áile Jávo, Sami, Samiland, Norway, COP21)

While national leaders all share responsibility for the climate, according to the substantive aspect of responsibility, this must take diverse forms in relation to diverse states (responsibility extends with extended power). The UNFCCC, ratified by 195 countries in 1992, presents the principle of "common but differentiated responsibility" (Klein, 2014, p. 410). What this means in concrete terms is at the core of the COP negotiations. Although it has been discussed for some time, citizens remain confused about who is responsible for the climate: is it the UN, national governments, the corporate world, rich people, or all individuals (Hulme, 2009, p. 318)? This is where choosing between the all-affected and all-subjected principles makes all the difference; namely whether you stipulate that responsibility should be assigned according to *effects* or according to *structures of governance*. Moreover, there remains a choice concerning how far responsibility should be extended in space. This brings us to the crucial consideration of geographical scales of justice.

The Geographical Framing of Justice: Re-theorizing Justice and the Media in a Globalizing Age

Rawls (1971, p. 118) elaborates on ideal conditions for justice with his concept of "the veil of ignorance". This assumes you know nothing about the context, or about yourself and your position in society, and thereby allegedly creates justice in the purest form. This runs contrary to postcolonial scholarship that aims to decolonize methodologies[5] that "veil" privilege, among other things. Take, for instance, privilege relating to the hegemony of the nation-state. Rawls furthermore states: "I shall be satisfied if it is possible to formulate a reasonable conception of justice for the basic structure of society conceived for the time being as a closed system isolated from other societies" (Rawls, 1971, p. 7). This closed system is the nation-state system where one state is the focus, not relations between states or transnational relations for that matter (Rawls, 1971). But what happens when an intranational justice problem arises within the boundaries of the nation-state? Take, for instance, the problem of how these very boundaries were drawn through colonization, as is the case for most indigenous groups? It is at this juncture that Rawls' (1971, pp. 10–11) idea of "justice as fairness", which stipulates that the principles of justice should be agreed upon from a "fair" point of departure, seems unattainable when formulated in relation to such internal conflicts. And the principle aiming to compensate for unfair opportunities, the "fair equality of opportunities" principle (Rawls, 1971, p. 73) is also designed and managed by the institutions of the colonizing state (see also Hayden, 2010, p. 358).

Fraser (2008, 2014) acknowledges these problems, pointing out that most previous theories conceive of justice as "normal justice" and construct normative principles only relevant in contexts where there is no significant disagreement on justice claims. For example, constructing these principles along the nation-state scale would mean overlooking "abnormal justice" that includes transborder injustices, which are highly relevant when regarding climate change, indigenous peoples and the media.

In a global age, attentiveness to the substance of justice and notions of responsibility must therefore be accompanied by attention to geographical justice scales (Fraser, 2008, 2014; see also Hayden, 2010). Distributive justice, for instance, can simultaneously function on the international and global levels (Caney, 2010a, 2010b). *Misframing* (Fraser, 2008), the worst injustice in a globalizing world, occurs when irrelevant geographic scales are imposed upon cases of injustice, such as when national scales and solutions are employed to address transnational sources of injustice.[6] This is highly relevant to climate change where causes of injustice stretch beyond national scales; with effects spreading, for instance, across several nations in the Polar Regions or in the Amazon rainforest. In terms of the substance of justice, misframing is connected to representation. There are at least two variants of misrepresentation in Fraser's (2008, p. 18) account: ordinary political misrepresentation, concerning how electoral systems are arranged and *misframing*, where the boundaries around a community are drawn up wrong, resulting in meta-injustice. In these scenarios, ultimately those constituted as non-members cannot claim redistribution, recognition or ordinary political representation. This is furthermore connected to a form of meta-political misrepresentation in the frame-setting processes that concerns who gets to set the frames and determine who gets to participate.

In tackling misframing, an *affirmative politics of framing* (Fraser, 2014) would mean looking beyond national scales in certain cases without questioning national framing as such. A *transformative politics of framing* (Fraser, 2014) looks further, questioning the use of national scales in remedies to transnational injustices in *any* case. The global justice movement and indigenous environmental activism groups have applied the latter, challenging the very structure of international politics by questioning the hegemony of the geographical scale of the nation in relation to the substance of justice. Hence, intersections of indigenous peoples and climate justice, and how indigenous peoples try to voice their experiences and opinions on this in the media, are well suited to help elaborate on theories of justice for a globalizing world. Fraser (2008, p. 51) writes that "claims for representation increasingly break the previous frame of the modern territorial state".[7] Such claims are made by many of the indigenous activists we interviewed (see Chapter 5). In Fraser's terminology, representation concerns *political* representation. However media representation is connected, since it is important "not only for democracy but for the pursuit of justice in general" (Sen, 2009, p. 254).

Parallel to Fraser's trajectory, Couldry (2012) discusses justice in relation to globalization as well as digitalization, identifying the need for a new media ethics (see also Deuze, 2005; Rao & Wassermann, 2015). Since the media play a crucial role in representing facts and norms that guide human action (Couldry, 2012), analyses of media globalization must decipher the interconnectedness of journalism, ethics, democracy and justice (Muhlmann, 2010; Rao, 2015). Couldry (2012, p. 181) discusses *environmental* ethics in relation to global environmental damage to demonstrate the relevance of what we, with Fraser (2008), would call a *global reframing*. As crucial framing agents in contemporary societies, media hold great potential to contribute to or even lead such re-framing efforts by making injustices visible:

> We need to push people *with information* to get them to act, and to stand up for climate justice, for the rights of ... indigenous peoples. (Jannie Staffansson, Sami, Samiland, Sweden, COP21, emphasis added)

Silverstone (2007, p. 144) underscores that as institutions media act justly and unjustly, for instance by "denying or distorting the voices of others." He identifies indigenous peoples as an example of this, together with marginalized and minority groups (which also often applies to indigenous populations). In a globalizing world, the severity of such foreclosing deepens with increases in the size of audiences who may be able to hear but cannot due to the systematic exclusion of such voices and perspectives (Silverstone, 2007).[8] In line with this, it is now time to delve deeper into climate justice along diverse geographical scales.

The Climate in Climate Justice

Since it has long been clear that the injection of carbon from the burning of fossil fuels (coal, oil, gas) causes dangerous climate change, climate justice discussions have largely concerned issues of how to stop or regulate carbon emissions (Shue, 2014, p. 1). In relation to this, three main questions can be formulated when general justice approaches are applied to the climate:

1. What needs to be done? What measures are necessary to tackle climate change?
2. Who should do it? Which type(s) of actors are relevant in the context of climate change?

3. How should this work be distributed? How should relevant actors divide the burden that climate measures necessitate? (Brülde & Duus-Otterström, 2015, p. 75)

Each of these questions evoke issues about what substance of justice should be emphasized: redistribution, recognition (identity/status) or representation (political). These scales are mainly salient for the questions regarding: what needs to be done, what measures are necessary to tackle climate change, and which types of actors are relevant in the context of climate change. The questions also raise issues concerning responsibility and geography, mainly by asking who should do it, how this work should be distributed, and how relevant actors should distribute the burden. In addition to Brülde and Duus-Otterström's questions, we argue, following Page (2012), it is important to consider who needs to do what *to achieve which goal* (see also Brülde & Duus-Otterström, 2015)? People may incidentally not agree on what we need to achieve nor on what is deemed a tolerable rise in temperature, if any.

The main measures discussed in relation to climate change are: mitigation, adaptation and compensation. According to Brülde and Duus-Otterström (2015, p. 76), all three are needed but there is a hierarchy to them. Mitigation should come first, adaptation should be used only in cases where mitigation is too late, and compensation should, in turn, be applied only when mitigation and adaptation both fail to tackle the situation. Compensation is costly and problematic, among other things because a person who is forced to leave his or her native environment cannot be compensated for the emotional loss (Brülde & Duus-Otterström, 2015, p. 77). Mikaele Maiava, a Tokelauan from Tokelau in the Pacific Ocean expresses how strongly people there feel about their native environment:

> … the people … are really strong on staying on the island and that means … if anything happened they will prefer to die with the island. So you are talking about … genocide … People won't move because that's who they are, that's where they belong. So, yes, it's really a huge impact to us and it will destroy everything that we live for. (Mikaele Maiava, Tokelauan, Tokelau, COP21)

Maiava advocates the mitigation measure: "We hope that they will have agreements … to keep the level under 1.5 degrees." This measure is not only the most effective for stopping or decreasing the negative effects of climate change; it is also considered the least costly alternative. Brülde and Duus-Otterström (2015, p. 78) identify *costly* measures needed to tackle climate change when discussing climate justice, not profitable ones, since profitable *and* effective measures do not pose ethical challenges and are thus not part of discussions on distributing the so-called climate burden. We consider these costly measures crucial for climate justice discussions. However, less costly measures are also brought to the fore in relation to recognition and representation (rather than redistribution).[9]

Understandings of climate change cannot be separated from how we understand development (see Hall, 2016). The way we define poverty and understand global inequality are crucial to how we understand climate change (Hulme, 2009, p. xxxvii; see also Brülde & Duus-Otterström, 2015, p. 82). Hence, it is imperative to distinguish between luxury emissions and survival emissions (Agarwal & Narain, 1991; Brülde & Duus-Otterström, 2015, p. 81; Shue, 2014). The primary emissions to be cut are luxury emissions, mainly those produced in the Western world. Survival emissions for the poor—such as the ability to store food, to access clean drinking water, to heat and cool your home, and to mechanize the heaviest labour—should not be targeted, at least not in the first step. However, as Shue (2014, p. 7) points out, it is not the right to emissions that needs to be guaranteed to poor people, but the right to energy. Energy can be retrieved in diverse ways, not only by fossil fuels.

Elaborating further on climate change's entanglement with how we normatively and sociologically understand our world, Hulme writes that:

> [C]limate change is not a problem that can be solved in the sense that, for example, technical and political resources were mobilised to 'solve' the problem of stratospheric ozone depletion. We need to approach the idea of climate change from a different vantage point. We need to reveal the creative *psychological*, *spiritual* and *ethical* work that climate change can do, and is doing for us. (Hulme, 2009, p. xxxvii, emphasis added)

This can be done by discussing climate change in terms of climate justice, which evokes the *ethical* aspect; and by drawing on diverging views on nature, such as indigenous peoples' views on Traditional Ecological Knowledge (TEK) (Guttorm, 2011; McGregor, 2002, 2004), which evokes the *spiritual* aspect.[10] Furthermore, we may *psychologically* embrace a Common Human Identity (CHI).[11] It is however important to concurrently note that:

> … the powers of Europe, with their early industrialization and imperial histories, bring a different historical legacy into climate negotiations than do their former colonies, now emancipated and seeking redress for past injustices. (Hulme, 2009, p. 318)

Climate change can thereby also be discussed in terms of a *political ideology*. It is caused by emissions that are manifestations of a "free-market, consumption-driven, capitalist, economy" (Hulme, 2009, p. xxx; Klein, 2014). This is something indigenous climate justice activists alluded to at times. Francois Paulette, a former Chief of Fort Smith First Nation, for instance, connects destructive free-market, consumption-driven, capitalist environmental practices and ideology to *media* practices and ideology.

> They are increasing the tar sands output which will destroy more environment … the river, the people … most of the papers in the world are right wing conservative papers so they reflect the profile of industry and governments rather than the people. (Francois Paulette, Dene people, Canada, COP17)

Among the diverse approaches the mainstream media can take towards climate justice, Schmidt and Schäfer (2015) identify five ideal types in mainstream quality print media in Germany, India and the USA. These ideal types are: freedom and resilience of peoples; fairness and effectiveness through market mechanisms; provident global government for people and the planet; international solidarity in an unequal world; and economic growth and social justice. Their climate justice trajectory is generally characterized by a national level focus (Schmidt & Schäfer, 2015, p. 537). This nation-state focus sometimes encompasses intra-national dimensions, most often comprises international ones, and sometimes includes transnational or global aspects. In the following, we discuss diverging geographical framings of climate justice. We do not discuss the national level as an autonomous level, but view it as constituted by and constituting either intra-national or international relations (because how can a nation be defined except through its internal and external power-relations?). We start by detailing the international level, which is the dominant geographical scale in climate justice research.

International Climate Justice

Perspectives on international climate justice permeate the main controversies at UN COP meetings. In 2013, for instance, the G77 nations and China walked out of negotiations at COP19 in Warsaw to protest the lack of compensation for developing countries facing extreme climate events; a measure opposed by the EU, Australia, the United States and other "developed countries". As weather extremes increase, acute climate change effects will be worse in "developing countries" (Reese, 2016, p. 521; Santarius, 2009).

International climate justice moreover concerns "… what kinds of allocations of costs between rich nations and poor nations it would be acceptable for a poor nation to accept" when climate change is tackled (Shue, 2014, p. 27).

Schmidt and Schäfer (2015, p. 537) deem the national level of climate justice most important, specifically concerning how it is negotiated by governments *on the international level*. Rhaman (2014) has analyzed how Bangladeshi journalism approached climate justice in coverage of three UN climate summits (Bali, 2007; Copenhagen, 2009; Durban, 2011). He concludes that climate justice is mainly framed as an international issue, while noting that the national level (i.e. intranational level) must be addressed more in journalism to distinguish not just between rich and poor internationally, but also intranationally. Rhaman also emphasizes the importance of journalism that targets the politics and mechanisms behind the summit process itself when discussing climate justice (a matter we shall return to in Chapter 5).

While the paths chosen may differ, three questions have guided discussions on responsibility for the burden of distribution when tackling climate change: (1) Who has contributed most to the problem? (2) Who has benefitted most from (the causes of) the problem? (3) Who has the greatest ability to solve the problem? (Brülde & Duus-Otterström, 2015, p. 86). Three solutions tend to emerge that each emphasize one of these questions more than the others: (i) the Polluter Pays Principle (PPP), (ii) the Beneficiary Pays Principle (BPP) and (iii) the Ability to Pay Principle (APP) (Brülde & Duus-Otterström, 2015, pp. 87, 93, 96). Each of these principles emphasizes who is to pay, implying that justice has to do with redistribution, underlining economic rather than cultural and political aspects, which would have instead implied a focus on recognition and representation. As we shall see, interviewees mainly bring up recognition and political representation in discussions about activism and climate politics (see Chapter 5). These aspects are also evoked in discussions about intra- and transnational climate justice. Given the dominant position of these three principles dictating who is to pay, in climate justice research as well as in climate politics, we discuss them in detail below, and relate them subsequently to indigenous peoples.

PPP, BPP and APP are generally connected to the nation-state level, and more particularly to international relations. PPP is well established in national as well as international law (Brülde & Duus-Otterström, 2015, p. 87). While it was originally intended to save governments and tax payers from paying for emissions made by private companies, it has gradually become a leading argument in international negotiations concerning how the climate burden should be distributed between states. In its basic form, the PPP reads "States should take responsibility for a share of the joint climate burden that corresponds to the amount of fossil fuel emissions they have caused" (Brülde & Duus-Otterström, 2015, p. 87).[12] If distinctions are made between emissions and polluting emissions, and between states and sufficiently rich states, PPP can be reformulated to account for the right each state should have to some emissions, while acknowledging that some (rich) states need to take more responsibility than others: sufficiently rich states should take on a share of the climate burden that corresponds to how much polluting emissions they have made (Brülde & Duus-Otterström, 2015, pp. 88–89).

There are a number of arguments against PPP. For instance, you cannot distinguish between emissions and harmful emissions if there is no agreement on what each state is allowed to emit (Brülde & Duus-Otterström, 2015, p. 89). To counter this problem emissions are sometimes counted per capita per state as non-polluting if the same per capita amount could be emitted by all states without causing dangerous climate change. Other arguments against PPP concern unknowing and dead polluters (Brülde & Duus-Otterström, 2015, p. 90). Regarding unknowing polluters, lack of knowledge is only relevant if it is not caused by a refusal to look for information; it is only acceptable if it is excusable or irreproachable.

A conundrum in these discussions is, yet again, where to draw the line in time concerning when it is reasonable to presume that people knew (Brülde & Duus-Otterström, 2015, p. 88). As mentioned in Chapter 1, there is evidence that knowledge of climate change was possible as early as the nineteenth century, and at least since the 1950s (Brülde & Duus-Otterström, 2015, pp. 90–91). *The Limits to Growth* report was published by the Club of Rome think tank in 1972 (Hulme, 2009, p. 260). It was the first major study of global environmental issues (Urry, 2011, p. 165). The report stated that if population growth, industrialization, emissions, food production and resource depletion carried on, the limits to growth would be reached within a century (cf. Urry, 2011, pp. 165–166). The line is most commonly drawn in 1990, when the IPCC published its First Assessment Report with the expressed aim of authoritatively informing the world's states of its research findings (Brülde & Duus-Otterström, 2015, pp. 90–91). If the principle about strict responsibility (see above) is applied, actors should however be seen as responsible even if they could not know for sure that their actions would have negative effects. According to this view, since some of those responsible are dead, and since the dead cannot pay, part of the climate burden cannot be carried in a just way. Others have to carry it for the dead. This is assumed to be done by states, which exist even if people within their borders change (Brülde & Duus-Otterström, 2015, p. 92). The latter informs the argument for focusing on states in climate justice discussions.

The Beneficiary Pays Principle was developed to tackle the problem of who takes responsibility for the cost of emissions caused by people who are dead, or who could not know about the negative effects of their actions (Brülde & Duus-Otterström, 2015, p. 94). The BPP reads: States shall carry a part of the joint burden which corresponds to how much they have benefitted from the activities that caused the emission of greenhouse gases (Brülde & Duus-Otterström, 2015, p. 93). Ultimately, the same states that have emitted the most have also often benefitted the most. Ed Page (2012) defends the BPP, and pinpoints the burden in terms of: states benefiting most from activities causing climate change should bear the greatest burden of costs associated with preventing dangerous climate change. The principle is sometimes reformulated to only include the states that have benefitted most, resembling reformulations of the PPP (Brülde & Duus-Otterström, 2015, p. 94). A crucial distinction is naturally when a state began benefitting: how can the international community deal with "the fast-rising economies of the Global South—with China, India, Brazil and South Africa leading the pack—that are mostly responsible for the surge in emissions in recent years, which is why we are racing toward tipping points far more quickly than anticipated" (Klein, 2014, p. 412). Another modification of BPP is that the costs it would impose on a state to act cannot be higher than the benefits received (Brülde & Duus-Otterström, 2015, p. 94). Even with these additions the BPP is subject to criticism like the PPP, namely that dead and unknowing people constitute problems (Brülde & Duus-Otterström, 2015, p. 95). We should not, it is argued, pay for the benefits received by others (i.e. now dead). Another problem concerns the meaning of "benefit": benefit in relation to what? Is this a counterfactual issue? That is very hard to measure. If it is a counterfactual issue, what is measured is the factual production connected to emissions. This makes the BPP argument come very close to the APP argument (Brülde & Duus-Otterström, 2015, p. 96).

The Ability to Pay Principle reads: The total climate burden should be distributed according to the ability diverse states have to carry the relevant costs (Brülde & Duus-Otterström, 2015, p. 96). Ability is measured by the size of national income or wealth. An alternative version instead emphasizes *relative* ability, based on how much of a sacrifice it would be to pay. The fact that some states are accustomed to a fossil fuel intensive lifestyle speaks against this, since it makes their relative ability small (Brülde & Duus-Otterström, 2015, p. 97). This seems however to be dependent on the idea of relative ability being determined by *experienced* ability, rather than for instance distinguishing between luxury and survival emissions, which we argue could be done in connection with APP. The APP is an ahistorical principle, as it focuses on the ability to pay *now*, regardless of history. This is the most common argument against it. In the end, however, it is often those with the greatest ability to pay today that emitted most in the past (Brülde & Duus-Otterström, 2015).

Another argument against the APP is that it may discourage states shifting from fossil fuels to clean energy sources, since this positive move may not necessarily result in them paying less for the total damage (their ability to pay may still be considerable). People who support the APP argue that there does not need to be a debt or a blame to distribute in order for us to determine who is responsible, similar to the way people take responsibility and save someone who is drowning (Brülde & Duus-Otterström, 2015, p. 98). Like the drowning metaphor, it is simply the one able to help who should help. It is sometimes underlined that no one is to blame since the climate problem has not been predicable enough, and that the threat posed by the changing climate is joint and undeserved (Brülde & Duus-Otterström, 2015, p. 98). Nevertheless, given arguments from the other principles there is at least some agreement that we have known about climate change and damaging emissions for some time, and thus damage cannot be seen as completely undeserved by states. The bearing argument of APP proponents is that rich states should pay; they deserve to step up, as evidence used by PPP and BPP advocates shows. Indigenous peoples cannot be considered significantly responsible according to any of these principles. They have not polluted the most, they have not benefitted the most and they do not have the best ability to shoulder the burden. On the contrary, they could use all these arguments against the states within whose borders they dwell.

In international politics, most of the time the three principles regarding the dissemination of the burden between states will point to the same rich states, which have emitted and benefitted the most (see Brülde & Duus-Otterström, 2015, p. 99). In any case, most philosophers agree that none of the principles works for climate change without support from the others. Caney (2010c) takes the PPP as a starting point and emphasizes its traits while developing a theory encompassing all principles (Brülde & Duus-Otterström, 2015, p. 100). Caney combines PPP and APP, with BPP becoming integrated into a widened view of APP (Brülde & Duus-Otterström, 2015, p. 103). That is, most responsibility falls upon the major polluters, but only if they are able to pay.

Some controversy remains concerning which states should be considered unable to pay; what is the reasonable poverty limit? The emissions made by people who are dead or who were unknowing are, in Caney's version, termed "the remainder" which should, according to him, be shouldered by the rich states (here the APP kicks in) (Brülde & Duus-Otterström, 2015, p. 101). This is not considered completely just, but is seen as the least unjust possible option. Caney (2010c, see also Brülde & Duus-Otterström, 2015, p. 101) states moreover that the duty to do something about "the remainder" falls mostly upon the rich states *that achieved their wealth unfairly*. This way counter-arguments from rich states who want to avoid paying can be overcome. Common arguments put forward by rich states to avoid paying are generally: (1) it threatens economic growth, (2) it undermines their freedom, and (3) it is unfair because they have the right to their possessions (Brülde & Duus-Otterström, 2015, p. 102). The first argument may be valid, according to Brülde and Duus-Otterström, but the third one can be countered by a modified version of APP, stating that those who achieved their possessions unfairly cannot claim fairness in protecting them. As we shall see, this argumentation is used by indigenous peoples in protesting climate change and land rights issues concerning colonized lands on the intra- and transnational levels simultaneously.

Intranational Climate Justice

> Climate justice, I think it needs to have a strong link to those communities that are on the front lines of climate change, and the … differential impacts upon those people because of that, marginalized peoples, indigenous peoples, people … living in vulnerable environments … and recognizing that it's through colonization we have gotten into this … situation. (Teanau Tuiono, Māori, New Zealand and Cook Islands, COP21)

Intranational climate injustice is comparable to environmental racism; that is the practice of directing pollution and waste towards poor, non-white communities within a certain nation-state (Bullard, 1990). The discussion has recently been extended to include how *climate change* unevenly affects black people (Milman, 2016) and indigenous communities (Roosvall & Tegelberg, 2015; Tauli-Corpuz, & de Chavez, 2009). Teanau Tuiono (quoted above) articulates the specific impact of colonization in this regard, acknowledging how certain communities, colonized by nation-states and subsequently situated within them, are made particularly vulnerable. To some extent such injustice may stay unnoticed if affected groups are too marginalized to voice their opinions widely. Media logics may contribute to continued marginalization by excluding such voices, or by including such voices only in de-politicized contexts (Roosvall & Tegelberg, 2012, 2013, 2015). The "Economic growth and social justice" discourse—one of the ideal types for climate justice reporting identified by Schmidt and Schäfer (2015)—however, sometimes focuses on injustice at the intra-national level. By underlining, for instance, that measures must be taken for local reasons and domestic communities rather than for broader international reasons or communities (although the same discourse is sometimes also constituted on the international level).

Often when intranational environmental and climate justice activism concerns indigenous rights, it is simultaneously transnational. For example, a local movement in Gállok (the Samiland for an area known in Swedish as Kallak), in the Swedish part of Sapmi (Samiland), opposing mining that threatens the environment and indigenous rights is at the same time a global struggle (Cocq, 2016, p. 54). The mining planned in the area was to be conducted by a British company, Beowulf mining. During a presentation of the project, Beowulf mining's President, Clive Sinclair-Poulton, asked "What local people?", basically questioning the existence of any local people in the mining area. This question was re-appropriated by the local movement by creating the website www.whatlocalpeople.se (see Cocq, 2016, p. 56). On the website, local people are presented under the heading "We are the locals!" (Cocq, 2016, p. 57). Here an intranational aspect of justice is implied. It does not concern all of Sweden in the same way, even though it may be seen as threatening democracy in the nation to encompass such injustice without amending it. Instead it affects a specific locality or region, which, as is often the case with intranational injustice, affects marginalized groups more than others. Dahlberg-Grundberg and Örestig (2016) conclude that media users in this case were able to extend a locally and physically situated protest by linking it to a globally contentious issue such as the mining boom and its consequences for indigenous populations.

In *This Changes Everything*, Naomi Klein (2014) delineates in a similar way how local struggles against pipelines in North America connect to intranational justice, while concurrently encompassing transnational aspects. Klein (2014, p. 369) identifies the opponents to indigenous peoples in these struggles as government, industry, police and corporate-owned media. She stresses that these media not only portray indigenous peoples as "living in the past and enjoying underserved special rights", but also fail to "do basic public education about the nature of the treaties our governments (or rather their British predecessors) had signed" (Klein, 2014, pp. 369–370). Vibeke Larsen, a Sami politician in Norway recognizes the latter as a major problem concerning media reporting on indigenous people:

> The media don't know the indigenous peoples … it seems like a lack of knowledge. (Vibeke Larsen, Sami, Samiland, Norway, COP17)

That is, national media do not know the indigenous peoples living within the borders of the nation-state. At the same time, indigenous rights have gradually come to play a central role in fossil fuel resistance, with some non-indigenous people realizing what powerful tools they can be against ecological crisis and how much indigenous ways of life can teach all people about "how to relate to land in ways that are not purely extractive" (Klein, 2014, p. 370). In Canada, many recent land rights cases have been acknowledged as shared by settlers and First Nations. Yet indigenous groups cannot properly exercise these land rights if the other party is irrevocably altering and poisoning the land they are supposed to share (Klein, 2014, pp. 371–372). This acknowledgement has allowed indigenous fishermen to fish outside of government-approved seasons, which has led non-indigenous fishermen to attack them, sinking their fishing boats, hospitalizing several indigenous persons (Klein, 2014, p. 373). Klein (2014, p. 373) underlines that for a time the racism was so fierce that a "non-Native fisherman put on a long-haired wig and performed a cartoonish 'war dance' on the deck of his boat in front of delighted television crews".

Hence, recent efforts to highlight indigenous rights and their connection to climate and environmental struggles have, on one hand, caused understanding and co-operation between indigenous and non-indigenous peoples. On the other hand, they have led to violent attacks by non-indigenous peoples on indigenous peoples. Mainstream media seem to have thrived on this racism, as exemplified above ("delighted television crews"). In contrast, some media have reported on people coming together in these struggles, as this Al Jazeera English report from New Brunswick, Canada shows:

> "This is not just a First Nations campaign. It's actually quite a historic moment where all the major peoples of this province—English, French and Aboriginal—come together for a common cause," David Coon, head of the Green Party in New Brunswick, told Al Jazeera. "This is really a question of justice. They want to protect their common lands, water and air from destruction." (Lewis, 2013; cf. Klein, 2014, p. 374)

However national law does not yet protect indigenous peoples when they oppose extractive projects that are obviously illegal. On the contrary, indigenous peoples often find themselves "on the wrong side of a can of pepper spray—or the barrel of a gun" (Klein, 2014, pp. 377–378). In fact, according to Global Witness, in 2015 alone 185 environmental activists were murdered worldwide, two-thirds from Latin America (Salazar, 2016). The murder of indigenous human rights defender Berta Cáceres in her home in Honduras was noticed around the world; while too many others remain unreported on a larger scale. As Klein notes, "… isolated, often impoverished Indigenous peoples generally lack the monetary resources and social clout to enforce their rights, and anyway, the police are controlled by the state" (Klein, 2014, p. 378).

Nation-state governments have benefitted from the injustice inherent in the gap between rights and resources when it comes to indigenous peoples (see Klein, 2014, p. 380). In this context, climate change is furthering the economic pressure on indigenous communities to make deals with the fossil fuel industries (Klein, 2014, p. 385). Hence, the opportunities for co-operation and efforts to save the planet facilitated by the indigenous rights described above are countered by pressures felt within indigenous communities to make fossil fuel deals for economic survival.

> What we cannot expect is that the people least responsible for this crisis will foot all, or even most, of the bill. Because that is a recipe for catastrophic amounts of carbon ending up in our common atmosphere. Like the call to honour our treaties and other land-sharing agreements with indigenous peoples, climate change is once again forcing us to look at how injustices that many assumed were safely buried in the past are shaping our shared vulnerability to global climate collapse. (Klein, 2014, p. 418)

The injustices described here are partly intranational, they are matters between colonizing states and colonized indigenous peoples. Brülde and Duus-Otterström (2015, p. 83) claim that nation-states are the most responsible for acting on climate change. They provide three main reasons for this, each related to the intranational geographical scale: (1) States have jurisdiction over their own territories (even if corporations have been responsible for certain emissions, the state is ultimately responsible for emissions done in its own territory, since it could have regulated or stopped this); (2) Emissions produced by the public sector are often huge (while individuals enjoy the benefits of these emissions, they cannot be held responsible for them since they are determined by the state); (3) States have a unique capacity to make a difference (they can coordinate and encourage large-scale action) (Brülde &

Duus-Otterström, 2015, pp. 84–85). The third argument also encompasses the first one (the state jurisdiction), but while the first one emphasizes retrospective responsibility, the third underlines responsibility for the future.[13]

Anthony Debbarma of the Borok indigenous peoples, relates to the intranational responsibility of states as follows:

> … let me just cite the Indian constitution … every individual has a right to protect the environment, the forest, the rivers, the ponds, the wildlife, animals, all this. If … the legal framework that has been laid down universally is not being respected, then of course climate justice is not there. (Anthony Debbarma, Borok people, India, COP21)

Debbarma refers to the Indian constitution in particular *and* to universal rights. This once again underlines the fact that the local, intranational struggles of indigenous peoples are often at the same time universal. This universality, the *global* nature of the climate crisis, which operates at a *transnational* scale, can sometimes result in the strengthening of *regional* identities. This can, in turn, lead to concern for *local* justice, regional efforts to amend injustice and to nature conservation efforts as well (Reese, 2016, p. 526). Thereby the sense of a global climate change crisis may help raise consciousness for intranational climate justice. We elaborate on how a sense of globality can be evoked in the next section.

Transnational and Global Climate Justice

Global climate change is an unprecedented challenge for all humanity; "responding to its consequences requires exceptional *transnational* efforts that have to take into account the *inequalities* fostered by climate change" (Reese, 2016, p. 529, emphasis added). Hence what is needed is transnational climate justice. Increasingly researchers suggest that we must include psychological aspects when discussing how to tackle climate change (Hulme, 2009, p. xxxvii), and specifically regarding its transnational aspects (Reese, 2016). In re-situating culture and humans at the core of our understanding of the climate, Hulme (2009, pp. xxxvii–xxxviii) states that "[h]uman beings are more than material objects, and climate is more than a physical entity … the idea of climate change is an imaginative recourse around which our collective and personal identities and projects can, and should, take place."

In the same spirit, Reese proposes that social identification with all humans may represent a potential path to global environmental justice, since, in general, higher identification with a superordinate group (all of humanity) leads to greater focus on common goods (Huo, Smith, Tyler, & Lind, 1996; cf. Reese, 2016, p. 527). This social identification with all humans is reminiscent of Young's (2013) social connection model. As previously mentioned, this model describes how responsibility for justice must transgress national borders through social connection. Reese elaborates on psychological aspects of this: "When individuals identify with a relevant group, they perceive themselves as components of a higher order social unit" (Reese, 2016, p. 523). More specifically, a common in group identity model (Gaertner & Dovidio, 2014) postulates that by regrouping smaller groups into a common superordinate group, intergroup relations can improve and pro-social behaviour can be fostered (Reese, 2016, p. 523). Reese also refers to the *subject* of environmental justice, stating that one way to deal with the global climate crisis is to focus on a common bond between all humans; a Common Human Identity (CHI), *where the group of all humans becomes the subject of environmental justice* (Reese, 2016, p. 527; Sikor, Martin, Fisher, & He, 2014). This connects to the all-subjected principle of justice as formulated by Fraser (2008, 2014) and Owen (2014). However, in Reese's version, people are subjected by their common human identity, by being human.

Experiments have shown that the more people identified with a CHI, the more they believed that natural resources should be used for basic needs rather than for material wealth (Reese, 2016; Reysen & Katzarska-Miller, 2013). This would mean that a CHI could help emphasize survival emissions above luxury emissions (see Shue, 2014). A World Value Survey (WVS, waves 5 & 6, 2005–2014), a representative study accounting for around 90% of the world's population, moreover indicated that the more people identify as global citizens the more they are concerned with pro-environmentalism (via Rosenmann, Reese, & Cameron, 2016, cited in Reese, 2016, p. 526). Hayden (2010, p. 368) likewise describes how a world environmental citizenship arises from "an ethical concern for the social, political and economic problems associated with the environment and humanity's dependence upon it, and from a recognition of our global responsibilities for the human condition in light of humanity's interconnectedness with the environment." While correlating positively with concern for the natural environment, CHI also correlates negatively with anthropocentrism (Reese, 2016, p. 526). This corresponds to the holistic worldviews and the environmental engagement expressed by many of the indigenous activists we interviewed and many indigenous groups in general (see e.g. Adams, 2004; Apffel-Marglin, 1998; Guttorm, 2011; and see Chapter 3).

Moreover, when Reese specifies the work needed to be done to form a CHI he stresses that subgroups can be formed that formulate relevant shared goals, which subsequently end up in an overarching group representing all subgroups who are then able to formulate shared goals on a higher and broader level, based on the consensus of the goals of the subgroups (Reese, 2016, p. 527).

This is something that seems to characterize the co-operation between indigenous groups from all parts of the world. They stand side by side at the COPs and express solidarity with each other across continents. Take, for example, expressions of solidarity from Sami in Scandinavia to First Nations in Canada during the highpoint of the Idle No More Movement, or from groups in the US to the Sami in Sweden during the struggle against environmentally destructive mining on traditional Sami territory, and more recently from indigenous groups around the world who supported the NoDAPL movement protesting against a pipeline at Standing Rock, North Dakota in 2016. In fact, the work of indigenous groups materializes on a broader and more integrated level than those imagined by Reese. While Reese (2016, p. 526) sees a local and a global identity as providing *two pathways* to justice (as well as to conservation) on the local and global level, indigenous groups combine these paths by emphasizing both intra-national and global problems simultaneously. They also do so by working together regionally—within and across nations—and globally, as illustrated by the events organized at the Indigenous Peoples' Pavilion as well as in protest actions and testimonies from indigenous peoples.[14]

Jannie Staffansson says:

> … if people, if politicians and the private sector would have used the term climate justice the way I refer to it as also including the indigenous peoples and respecting our rights to territories and land, that would be very important and it would change my daily life because a lot of my time and a lot of our Sami people's time is going into trying to protect our land from exploitation. (Jannie Staffansson, Sami, Samiland, Sweden, COP21)

Staffansson implies here that she does not think the term climate justice is used pertinently by politicians and the private sector. She recognizes climate justice issues as intra- and transnational; the Sami are transnational people in intranational settings. Staffansson relates this situation to the struggles of global indigenous peoples more generally.

While transnationalism can be hard to grasp, it is admittedly harder to confine nature in space and time in accordance with the borders of nations-states and other limiting bodies. Klein (2014, p. 378) writes that "… while the lawyers argue the intricacies of land title in court, buzzing chainsaws proceed to topple trees that are four times as old as our countries, and toxic fracking seeps into the ground water". The trees were there before the borders were drawn, and the ground water of course knows no nation-state or other boundaries. We all share the same sky, a global atmosphere (Hulme, 2009, p. 319), an atmospheric commons (Klein, 2014, p. 388). With this comes a shared future, as underlined in the title of the famous 1987 Brundtland Report from the World Commission on Environment and Development: "Our Common Future" (see Hulme, 2009, p. 248). Reese's ideas on a CHI as well as Klein's and Hulme's takes on the atmospheric commons, can be related to what has been termed a transnational *media* commons (Eide, 2009). The term refers to a "mediated common grounds". How can such a grounds be constituted? Reese (2016, p. 528) focuses on possible symbols (equivalent to national flags or religious symbols) when discussing the importance of communication for the creation of a Common Human Identity. This can be taken further by going beyond simple emblems and looking into communication in a broader sense by exploring how *ideas on humanity* are represented in the media. We elaborate on this in the concluding section of this chapter.

Conclusion

In terms of ideas on humanity, Lilie Chouliaraki (2013, p. 26) suggests an attitude towards others constituted by a cosmopolitan solidarity (see also Fine, 2012). This entails a disposition to act towards vulnerable others across the world, without the anticipation of reciprocation. She notes that mediated communication is imperative in constituting such relationships with people that we will, in most cases, never meet face-to-face (see also Silverstone, 2007).[15] Common ground certainly can (and needs to) include the recognition of different conditions; namely that common ground does not rest on the idea that we have common conditions, but rather that every human shares the universality of vulnerability, sometimes materialized, sometimes not (Chouliaraki, 2013, p. 26).

Olga Atsynga Letykai Csonka underlines shared as well as specific experiences, with intra- and transnational dimensions:

> … if they open the Northern Sea Road along the Russian Arctic Coast it will be … a different time. Because … the animals go through the Bering Strait, and, of course,

> it is also the question for Alaska. So, indigenous communities from Alaska and Chukotka, we closely work together and hope Kamchatka will join us as well … So, we—now we understand what climate justice means for Russian indigenous peoples. (Olga Atsynga Letykai Csonka, Chukchi people, Russia, COP21)

What climate justice means for Russian indigenous peoples is apparently something concurrently intra- and transnational, shared and specific. Could this duality be a problem in relation to a CHI? A Common Human Identity connects to ideas on globalization as universalism rather than globalization as multiculturalism; to ideas on humanity as one rather than as constituted of particular(-istic) groups (Roosvall, 2016). Could universalism hurt indigenous peoples' particularistic claims, for instance, regarding affirmative action and restitution of previous wrongs? Findings from Reese's study counteract this caveat; stronger CHI predicts "beliefs that global socio-economic inequalities are unjust: The more respondents identified with a CHI, the more they believed that the inequalities between the rich and the poor in the world were not justified, but unfair and illegitimate" (Reese, 2016, p. 525; see also Reese, Proch, & Cohrs, 2014). A similar approach can be seen among the indigenous groups that identify with a CHI (identity) and recognize global—as well as intra-national—socio-economic inequalities (status).[16] As Seyla Benhabib (2006) notes, the tensions between the universal and the particular, the cosmopolitan and the local, are hugely important for our political future, and need to be elaborated on theoretically and analytically as a global network of institutions becomes amplified.

Reese concludes that CHI is something that can complement international policy-making; thus he includes the international scale as well as the intranational scale in the global scope of this identity. By building on the CHI idea, Reese (2016, p. 527) claims the implications for policy-making are that all nations should provide resources for the creation of subgroups *within each nation*; subsequently the interests of these subgroups can be combined along with those of different nations in order to form a superordinate group framework. This would constitute a trajectory from the local intra-national level, via the national, and then from the international to a transnational, global level. We note, however, that it is important that justice seeps back down to the local level, and that this level is not overshadowed by the more large-scale ones. Concern for the global is not really concern for the whole world if some "remote" communities constituted by minorities, such as indigenous peoples, are left behind, or used only as intermediaries to help save other "less remote" parts of the world (as sometimes happens in the media, see Chapters 4 and 5).

Therefore, discussions of the PPP, APP and BPP should be related also to intra- and transnational scales. We argue however that it is imperative to not just transfer perspectives from international climate justice to the other scales, since these perspectives are often afflicted with a one-sided emphasis on (re) distribution. This overlooks recognition and representation, fails to discuss responsibility in relation to the "whats"/substances/matters of justice, and misframes justice and responsibility geographically. The next chapter therefore delves deeper into the political geographies of climate change, in terms of the experiences of climate change in indigenous communities worldwide, as well as within the cartographies of climate summits and the connected activism there.

Returning to the questions set out in the introduction to this chapter, the most just ways to understand and approach issues of climate change in globalizing and unequal times include a combined focus on the *substance* and the *geography* of justice. Thus encompassing issues of the "what" (economy/culture/politics) as well as the who (national citizens/transnational groups, etc.) of justice. A focus on indigenous peoples and climate justice helps us develop this trajectory throughout the book, while illuminating the unjust situations indigenous people find themselves in regarding climate change issues, as well as rights and political representation in general. In connection to this, the most constructive ways to understand and approach *responsibility* for justice regarding the transnational phenomenon of climate change acknowledges that responsibility must also be transnational in scope.

We argue that the all-affected principle is thereby not enough. It must work in conjunction with the all-subjected principle, with its focus on structures of governance, in order to tackle the issue of unjust political *representation*, which is needed for misframing to be halted, and for the other "whats" of justice—*redistribution* and *recognition*—to be tackled effectively. The media can play a crucial role in tackling misframing since they are core framing agents in contemporary public discourse, setting up geographical as well as issue-related frames for every (climate change/climate justice) story told. Lastly, it is important to note that, in the climate justice literature and in politics, issues concerning intra- and transnational groups, and more particularly indigenous peoples, have been suppressed by emphasis on the dominant international geographical scale. The same can be said of media framings (see Chapters 4 and 5). Hence we argue the intra- and transnational dimensions must be better integrated into climate change and climate justice discussions—in politics, academia and the media, for instance, by using a Common Human

Identity approach that spans the local-global continuum and includes inter- as well as transnational dimensions. By crafting such an approach that pays heed to inequalities regarding status, we stand to gain a much-needed attentiveness to the complexities of justice in a globalizing world, and to how this might be addressed acutely in the context of climate change.

Notes

1. For further discussion of media ecology, see Chapter 4, and for further discussion of access to politics for indigenous peoples, see Chapter 5.
2. For further discussion of the roles of indigenous peoples in this regard, see Chapter 5.
3. We elaborate further on this in Chapters 4 and 5. See also Roosvall and Tegelberg (2013).
4. Silverstone (2007) also distinguishes between individual and procedural responsibility.
5. We elaborated on this project in Chapter 1. See also Smith (1999).
6. Misframing can also be more closely connected to voice (see Chouliaraki, 2013), which we discuss further in Chapter 5.
7. For more on claims regarding rights across borders, see Benhabib (2009).
8. We elaborate on these concerns in Chapter 4 on media ecologies and in Chapter 5 on the political, voice and agency.
9. For a discussion of profitable measures, see Neale, Spence & Ytterstad (2015) and Ytterstad (2015).
10. For further discussion of indigenous traditional ecological knowledge, see Chapter 3.
11. We elaborate further on the Common Human Identity (CHI) in the section on transnational and global climate justice.
12. Quotes from Brülde & Duus-Otterström's text have been translated from Swedish.
13. The responsibility of states however does not exclude the responsibility of individuals within states (Brülde & Duus-Otterström, 2015, p. 86).
14. For further discussion of indigenous political actions, see Chapters 3 and 5.
15. We shall discuss solidarity further in Chapter 6.
16. We will elaborate on how this relates to political representation in Chapter 5. In Chapter 6 we will expand on how universalism and particularism can intersect in diverse ways with recognition of identity and status respectively.

References

Adams, M. (2004). Negotiating nature: Collaboration and conflict between aboriginal and conservation interests in New South Wales, Australia. *Australian Journal of Environmental Education*, *20*(1), 3–12.

Agarwal, A., & Narain, S. (1991). *Global warming in an unequal world*. New Delhi: Centre for Science and Environment.

Apffel-Marglin, F. (1998). *The spirit of regeneration: Andean culture confronting western notions of development*. London: Zed Books.

Arendt, H. (1998). *The human condition* (2nd ed.). Chicago, IL: University of Chicago Press.

Ayish, M. I. (2015). Justice as an Islamic journalistic value and goal. In S. Rao & H. Wasserman (Eds.), *Media ethics and justice in the age of globalization* (pp. 139–154). London: Palgrave Macmillan.

Benhabib, S. (2006). *Another cosmopolitanism: Hospitality, sovereignty, and democratic iterations*. Oxford: Oxford University Press.

Benhabib, S. (2009). Claiming rights across borders: International human rights and democratic sovereignty. *American Political Science Review, 103*(4), 691–704.

Berlin, I. (2013). Alleged relativism and eighteenth century European thought. In H. Hardy (Ed.), *The crooked timbre of humanity* (pp. 73–94). Princeton, NJ: Princeton University Press.

Boltanski, L., & Thévenot, L. (2006). *On justification*. Princeton, NJ: Princeton University Press.

Brülde, B., & Duus-Otterström, G. (2015). *Klimatetik: rättvisa, politik och individens ansvar* [Climate Ethics: Justice, politics, and the responsibility of the individual]. Stockholm: Thales.

Bullard, R. (1990). *Dumping in Dixie: Race, class, and environmental quality*. San Francisco, CA: Westview Press.

Caney, S. (2010a). International distributive justice. In G. W. Brown & D. Held (Eds.), *The cosmopolitanism reader*. Cambridge: Polity.

Caney, S. (2010b). Global distributive justice and the state. In G. W. Brown & D. Held (Eds.), *The cosmopolitanism reader* (pp. 134–150). Cambridge: Polity.

Caney, S. (2010c). Climate change and the duties of the advantaged. *Critical review of International Social and Political Philosophy, 13*(1), 203–228.

Chatterton, P., Featherstone, D., & Routledge, P. (2013). Articulating climate justice in Copenhagen: Antagonism, the commons, and solidarity. *Antipode, 45*(3), 602–620.

Chen, X. (2008). *Justice, humanity, and social toleration*. Lanham, MD: Lexington Books.

Chouliaraki, L. (2013). Mediating vulnerability: Cosmopolitanism and the public sphere. *Media, Culture and Society, 35*(1), 105–112.

Cocq, C. (2016). Exploitation or preservation? Your choice! Digital modes of expressing percptions of nature and land. In H. Graf (Ed.), *The environment in the age of the internet: Activists, communication and the digital landscape* (pp. 53–74). Cambridge: Open Book Publishers.

Couldry, N. (2012). *Media, society, world: Social theory and digital media practice*. Cambridge: Polity.

Dahlberg-Grundberg, M., & Örestig, J. (2016). Extending the local: Activist types and forms of social media use in the case of an anti-mining struggle. *Social Movement Studies*, 3, 1–14. Retrieved from http://dx.doi.org/10.1080/14742837.2016.1268955

de Onís, K. (2012). Looking both ways: Metaphor and the rhetorical alignment of intersectional climate justice and reproductive justice concerns. *Environmental Communication*, 6(3), 308–327.

Deuze, M. (2005). What is journalism? Professional identity and ideology of journalists reconsidered. *Journalism*, 6(4), 442–464.

Eide, E. (2009). Global divides and transnational media literacy. *Nordicom Review*, *29*(Jubilee Issue), 151–161.

Fine, R. (2012). The idea of cosmopolitan solidarity. In G. Delanty (Ed.), *Routledge handbook of cosmopolitanism studies* (pp. 376–386). Abingdon: Routledge.

Fraser, N. (2000). Rethinking recognition: Overcoming displacement and reification in cultural politics. *New Left Review*, *3*, 107–120.

Fraser, N. (2008). *Scales of justice: Reimagining political space in a globalizing world.* New York, NY: Columbia University Press.

Fraser, N. (2014). Transnationalizing the public sphere: On the legitimacy and efficiency of public opinion in a post-Westphalian world. In K. Nash (Ed.), *Transnationalizing the public sphere* (pp. 8–42). Cambridge: Polity.

Fraser, N., & Honneth, A. (2003). *Redistribution or recognition: A political-philosophical exchange.* London: Verso.

Fraser, N., & Olson, K. (Eds.). (2008). *Adding insult to injury: Nancy Fraser debates her critics.* London: Verso.

Gaertner, S. L., & Dovidio, J. F. (2014). *Reducing intergroup bias: The common ingroup identity model.* New York, NY: Psychology Press.

Guttorm, G. (2011). Árbediehtu (Sami Traditional Knowledge) as a concept and in practice. In J. Porsanger & G. Guttorm (Eds.), *Working with traditional knowledge: Communities, institutions, information systems, law and ethics* (pp. 59–76). Kautokeino: Norway Sámi allaskuvla.

Hall, N. (2016). *Displacement, development and climate change: International organizations moving beyond their mandates*. Abingdon: Routledge.

Hayden, P. (2010). The environment, global justice and world environmental citizenship. In G. W. Brown & D. Held (Eds.), *The cosmopolitanism reader* (pp. 351–372). Cambridge: Polity.

Hulme, M. (2009). *Why we disagree about climate change: Understanding controversy, inaction and opportunity*. Cambridge: Cambridge University Press.

Huo, Y. J., Smith, H. J., Tyler, T. R., & Lind, A. (1996). Superordinate identification, subgroup identification, and justice concerns: Is separatism the problem; is assimilation the answer? *Psychological Science*, *7*(1), 40–45.

Klein, N. (2014). *This changes everything: Capitalism vs. the climate*. New York, NY: Simon & Schuster.

LeBar, M., & Slote, M. (2016). Justice as a virtue. In E. N. Zalta (Ed.), *The Stanford encyclopaedia of philosophy*. Retrieved from https://plato.stanford.edu/archives/spr2016/entries/justice-virtue/

Lewis, R. (2013, October 21). Shale gas company loses bid to halt Canada protests. *Al Jazeera America*. Retrieved from http://america.aljazeera.com/articles/2013/10/21/shale-gas-companylosesbidforinjunctiontohaltcanadaprotests.html

McGregor, D. (2002). Traditional ecological knowledge and the Two-Row Wampum. *Biodiversity*, *3*(3), 8–9.

McGregor, D. (2004). Coming full circle: Indigenous knowledge, environment, and our future. *American Indian Quarterly*, *28*(3–4), 385–410.

Michaelova, A., & Michaelova, K. (2015). Do rapidly developing countries take up new responsibilities for climate change mitigation? *Climatic Change, 133*(3), 499–510.

Milman, O. (2016, July 24). World can't afford to silence us: Black church leaders address climate change. *The Guardian*. Retrieved from https://www.theguardian.com/environment/2016/jul/24/african-methodist-episcopal-church-climate-change-letter

Muhlmann, G. (2010). *Journalism for democracy*. New York, NY: Polity.

Nash, K. (Ed.). (2014). *Transnationalizing the public sphere*. Cambridge, UK: Polity Press.

Neale, J., Spence, T., & Ytterstad, A. (Eds.). (2015). *Global climate jobs*. Retrieved from https://globalclimatejobs.files.wordpress.com/2015/09/global-jobs-online-sept-2015.pdf

Nussbaum, M. C. (2013). Foreword. In I. M. Young (Ed.), *Responsibility for justice* (pp. ix–xxv). Oxford: Oxford University Press.

Owen, D. (2014). Dilemmas of inclusion: The all-effected principle, the all-subjected principle, and transnational public spheres. In K. Nash (Ed.), *Transnationalizing the public sphere* (pp. 112–128). Cambridge: Polity.

Page, E. (2006). *Climate change, justice and future generations*. Cheltenham, UK: Edward Elgar.

Page, E. (2012). Give it up for climate change: A defence of the beneficiary pays principle. *International Theory, 4*(2), 300–330.

Raboy, M., & Mawani, A. (2014). Are states still important? Reflections on the nexus between national and global media and communication policy. In C. Padovani, & A. Calabrese (Eds.), *Communication rights and social justice: Historical accounts of transnational mobilizations* (pp. 94–112). New York, NY: Palgrave Macmillan.

Rao, S. (2015). Practices of Indian journalism: Justice, ethics and globalization. In S. Rao & H. Wasserman (Eds.), *Media ethics and justice in the age of globalization* (pp. 121–138). London: Palgrave Macmillan.

Rao, S., & Wassermann, H. (Eds.). (2015). *Media ethics and justice in the age of globalization*. New York, NY: Palgrave Macmillan.

Rawls, J. (1971). *A theory of justice*. Harvard: Harvard University Press.

Reese, G. (2016). Common human identity and the path to global climate justice. *Climatic Change, 134*(4), 521–531.

Reese, G., Proch, J., & Cohrs, J. C. (2014). Individual differences in responses to global inequality. *Analyses of Social Issues and Public Policy, 14*(1), 217–238.

Reysen, S., & Katzarska-Miller, I. (2013). A model of global citizenship: Antecedents and outcomes. *International Journal of Psychology, 48*(5), 858–870.

Rhaman, M. (2014). Justice delayed-justice denied: Framing climate justice in Bangladeshi newspaper journalism during COPs. In B. León (Ed.), *Communicating climate change: From global agenda to media representation* (pp. 179–212). Salamanca: Comunicación Social.

Roosvall, A. (2016). Journalism, climate change, justice and solidarity: Exploring attitudes in editorials surrounding the IPCC reports. In E. Eide, R. Kunelius, M., Tegelberg, & D. Yagodin (Eds.), *Media and global climate knowledge: Climate journalism and the IPCC* (pp. 129–150). Basingstoke: Palgrave.

Roosvall, A., & Tegelberg, M. (2012). Misframing the messenger: Scales of justice, traditional ecological knowledge and media coverage of indigenous peoples and climate change. In E.

Eide & R. Kunelius (Eds.), *Media meets climate: The global challenge for journalism* (pp. 297–312). Gothenburg: Nordicom.

Roosvall, A., & Tegelberg, M. (2013). Framing climate change and indigenous peoples: Intermediaries of urgency, spirituality and de-nationalization. *International Communication Gazette*, *75*(4), 392–409.

Roosvall, A., & Tegelberg, M. (2015). Media and the geographies of climate justice: Indigenous peoples, nature and the geopolitics of climate change. *TripleC: Communication, Capitalism & Critique*, *13*(1), 39–54.

Rosenmann, A., Reese, G., & Cameron, J. E. (2016). Social identities in a globalized world: Challenges and opportunities for collective action. *Perspectives on Psychological Science*, *11*(2), 202–221.

Salazar, W. (2016, June 27). Activist deaths demand accountability. *AIDA/Aidaamericas.org*. Retrieved from http://www.aida-americas.org/blog/activist-deaths-demand-accountability?utm_content=buffer276e4&utm_medium=social&utm_source=twitter.com&utm_campaign=buffer

Santarius, T. (2009). Emissionhandel und globale gerichtigkeit [Emission trading and global justice]. In S. Böhler, D. Bongardt, & S. Frech (Eds.), *Jahrhundertproblem Klimawandel. Forschungsstand, Perspektiven, Lösungswege* [Climatic change as unprecedented problem: research, perspectives, solutions] (pp. 121–138). Schwalbach: Wochenschau Verlag.

Schmidt, A., & Schäfer, M. (2015). Constructions of climate justice in German, Indian and US media. *Climatic Change*, *133*(3), 535–549.

Sen, A. (2009). *The idea of justice*. Cambridge: Harvard University Press.

Shepard, P., & Corbin-Mark, C. (2009). Climate justice. *Environmental Justice*, *2*(4), 163–166.

Shue, H. (2014). *Climate justice: Vulnerability and protection*. Oxford: Oxford University Press.

Sikor, T., Martin, A., Fisher, J., & He, J. (2014). Toward an empirical analysis of justice in ecosystem governance. *Conservation Letters: A Journal of the Society for Conservation Biology*, *7*(6), 524–532.

Silverstone, R. (2007). *Media and morality: On the rise of the mediapolis*. Cambridge, UK: Polity.

Tauli-Corpuz, V., & de Chavez, R. (Eds.). (2009). *Guide on climate change and indigenous peoples*. Baguio City: Tebtebba Foundation.

Tokar, B. (2014). *Towards climate justice: Perspectives on the climate crisis and social change*. Porsgrunn: New Compass.

Urry, J. (2011). *Climate change and society*. London: Polity.

Young, I. M. (2013). *Responsibility for justice*. Oxford: Oxford University Press.

Ytterstad, A. (2015). Climate jobs as tipping point: Norwegian grassroots challenging the oil and climate change hegemony. In K. Borgnäs, T. Eskelinen, J. Perkiö, & R. Warlenius (Eds.), *The politics of ecosocialism: Transforming welfare* (pp. 163–180). London: Routledge.

· 3 ·

DIVERGING GEOGRAPHIES

Indigenous Peoples, Climate Change and the UN COP Summits

Interviewer: Do you use the term climate justice?

Lene Kielsen Holm: I have heard about climate justice. It is something … maybe I can tell a small part of our mythology, Inuit mythology.

Silla, or Silap Inua, that means the weather. Silla means also the human intelligence. So the weather and human intelligence are connected in our mythology or cosmology, we can say. Because silla … it is the same word, just saying the bigger silla, the bigger intelligence, that means the universe. So the universe, the weather of the globe, our weather … but also the human intelligence are connected.

In our stories, from many, many years, there is a woman below, down in the bottom of the sea. Her name is Satsuma Arnaa (Mother of the Deep). In Canada, they call her Sedna. She is the keeper of the environment, both the animals but also the land, and the water, and the ocean, and the air.

When Inuit have been misusing one of those … For instance, by shooting five seals, even if they know they only need two … the filth from this action comes to her hair. So her hair gets all the pollution, polluted.

Her anger can only get resolved once the Inuit shaman comes down to her to comb her hair and get the filth out of it.

And she will tell the shaman what it is that made her hair filthy and he will go back to the Inuit and tell them:

"You have done this incorrect thing to the environment and the animals. That's why she has been punishing you."

So the moral is that Inuit should live with a consciousness of sustainability. I think these two are connected.

Interviewer:	So, that is how you would define climate justice then with this kind of a cosmological connection?
Lene Kielsen Holm:	Yes, Yeah … in connection with our culture.
Interviewer:	So that concept already exists, it predates this wording for it?
Lene Kielsen Holm:	Yes. Before 1987, the first meeting with Gro Harlem Brundtland and 1992 [Earth Summit], many thousands of years before we were living with a consciousness of sustainable development.

—Lene Kielsen Holm, Inuit, Greenland, COP21

This story weaves together the themes addressed in this chapter on natural and political geographies of climate change. In Inuit mythology, environmental responsibility and climate justice are *interconnected* in the same way as weather (nature) and human intelligence. Albeit rooted in a particular geographic region (the Arctic), the moral of this story transcends the geographic boundaries of any particular culture or nation-state. Holm stresses that Inuit, whose traditional territory spans across four nations (Alaska—USA, Canada, Greenland and Russia), "should live with a consciousness of sustainability." Responsibility for the environment thus falls in the hands of a *collective*, rather than in the hands of any particular *individual.*[1] Moreover, in the recounting of this myth, environmental experience and human intelligence are intimately intertwined; an environmental consciousness born of the intimate ties between humans and their natural surroundings, and transmitted over many generations.

What comes across in Holm's story, and many of the testimonials in this chapter, is a lifestyle deeply rooted in nature as opposed to one that relates to nature primarily through consumption. This collective view of environmental responsibility stands in contrast to a dominant Western paradigm which places responsibility in the hands of individual consumers. Michael Maniates (2002, p. 45) terms this "the individualization of responsibility", recounting *The Lorax*, a famous Dr. Seuss story that warns of the dangers inherent in this narrowing of the environmental imagination (2002, p. 46). He sums up the crux of this argument as follows:

> When responsibility for environmental problems is individualized, there is little room to ponder institutions, the nature and exercise of political power, or ways of collectively changing the distribution of power and influence in society … Instead, the serious work of confronting the threatening socioenvironmental processes that *The Lorax* so ably illuminates falls to individuals, acting alone, usually as consumers. We are individualizing responsibility when we agonize over the paper-or-plastic choice at the checkout counter, knowing somehow that neither is right given the larger institutions and social structures. (Maniates, 2002, p. 45)

Dr. Seuss' tale concerning the pitfalls of individualized environmental responsibility overlaps with the rhetoric that has dominated negotiations at UN climate summits. Inside the summit halls, formal negotiations between nation-states, with varying degrees of influence over the process, stay fixated on developing abstract, market-friendly mechanisms for responding to climate change with little regard for *collective* demands for system change emanating from outside these highly regulated spaces. As Weisser and Muller-Mahn put it, "What was at stake [at COP21] were the positions within the system, but not the system as such" (2016, p. 15).

By contrast, when we asked indigenous activists what motivated their climate action, we heard repeated emphasis on a shared or collective sense of responsibility for the environment. Several interviewees identified with this collective environmental responsibility by referring to themselves as guardians or protectors of the natural ecosystems where they reside. For some, like Sami reindeer herder Jannie Staffansson the catalyst for climate action was deeply personal:

> I started to engage myself within the climate issue because I was forced into it. As indigenous peoples we are the guardians of nature and my family are reindeer herders. I am the guardian of the reindeers, of my reindeers, and I am very emotionally attached to those animals. They are so much of my identity. So seeing them hurt and seeing them being in a very fragile situation it forced me to try to register what's happened. And so it's due—it was from the beginning for the reindeers and the nature. That's why I'm fighting. (Jannie Staffansson, Sami, Samiland, Sweden, COP21)

Māori activist India Logan Riley also related climate action to guardianship, or *kaitiakitanga*. She explained her motivation to get engaged "… from a human rights angle, from an ecosystems view … [from] the Māori approach of kaitiakitanga and just being guardians of this planet and there for all the people and the environment." "Climate change," she added, "fits within the broader term of kaitiakitanga, or guardianship, and healthy relationship with the land. To be a good leader is to ensure that your community is healthy and the environment is healthy" (India Logan-Riley, Māori, New Zealand, COP21). As Borok spokesperson Anthony Debbarma put it, "everyone … has a right to protect the environment, the forest, the rivers, the ponds, the wildlife, animals, all this" (Anthony Debbarma, Borok peoples, India, COP21). These examples reveal some of the similar yet varying ways in which indigenous activists expressed collective responsibility for protecting the natural world.

The previous chapter defined and conceptualized climate justice in relation to transnational indigenous groups who are disproportionately impacted by climate change, and attend UN climate summits to advocate for remedies to this unjust situation. This chapter builds on that framework by describing how unequal impacts of climate change manifest geographically (Chatterton, Featherstone, & Routledge, 2013, p. 603) *in* frontline indigenous communities and *at* international climate summits. Our analysis is guided by two main questions: How do indigenous peoples relate to the issue of climate change and convey their experiences at different geographical scales? How is the geography of the summits constituted with regards to indigenous peoples? To address these questions, we begin by considering research in a range of disciplines that frames climate change as a problem of scale and, in some instances, seeks to overcome this problem by bringing indigenous knowledge, particularly traditional ecological knowledge (TEK), into conversation with Western research paradigms. Next we delve deeper into the geographical dimensions of climate change by looking more carefully at three distinct scales: climate change geographies where indigenous peoples live; spatial divisions *inside* UN summits; and climate justice actions *outside* the summits. The chapter combines thick description of our own participant observation with the words of interviewees to map these divergent geographies of climate change.[2] We conclude by discussing how these diverging spaces relate to each other in terms of power and the need for indigenous inclusion; particularly by enhancing their access to media coverage and political arenas.

Climate Change as a Problem of Scale: The Case of Indigenous Peoples

Knowledge of climate change is produced, circulated and valued on uneven geographical scales. As Mahony and Hulme (2016, p. 5) explain "… the global climate system dominates scientific and political discourses" at the expense of more localized, regionally specific knowledge of climate impacts. Consequently, researchers across a range of interdisciplinary fields have begun to frame climate change as a problem of scale. In science and technology studies, the problem of scale manifests itself as one of overcoming a spatial division between the *technoscientific* systems wherein climate knowledge is produced (laboratories, computer modelling systems, etc.) and the *subjective, experiential* dimensions of climate change that this abstract knowledge renders invisible (Callison, 2014; Hulme, 2010; Jasanoff, 2009; Mahony & Hulme, 2016; Orlove, Lazrus, Hovelsrud, & Giannini, 2014). In media and communication studies, the problem is framed as one of translating *global climate knowledge*, produced by scientific bodies like the IPCC, into discourses that are accessible to *local publics* (Kunelius & Eide, 2016; Mann, 2012; Moser, 2010; Pearce, 2010). In international politics, the scalar dynamic emerges to highlight tensions between highly regulated regimes of *international* climate governance (i.e. the UNFCCC process) and *transnational* climate justice activists articulating counter-narratives that call for "system change not climate change" (Chatterton, Featherstone, & Routledge, 2013; Klein, 2014; Roosvall & Tegelberg, 2015; Weisser & Muller-Mahn, 2016). Whether the scales are tilted in favour of the technical over the experiential or the international over the transnational, knowledge production, mediation and political deliberation concerning climate change have transpired on uneven geographical terrain.

These scalar divergences pose three interrelated challenges for indigenous communities on the frontlines of climate change.[3] The first challenge is overtly *geographical*. It relates to the uneven geographic scales upon which climate change impacts are experienced, and the vast distances between frontline indigenous communities and global centres of media and political power. Brugnach, Craps, and Dewulf (2017, p. 2) identify one of the direct consequences, noting that "… indigenous concerns and contributions with respect to climate change mitigation have played a very minor role in policy developments and decision-making processes supporting mitigation efforts". They argue, in turn, that this has made the implementation of global mitigation measures difficult in local contexts where they are viewed with suspicion; and have, at times, consequently had the adverse effect of exacerbating climate-related issues (Brugnach et al., 2017).

A second, interrelated challenge is *political*. It concerns the just representation of indigenous peoples in global climate governance. Here the tension lies between the international governance structure of the UNFCCC and the transnational identities of indigenous groups and other non-nation-state actors (Schroeder & Lovell, 2012). For example, in northern Europe,

Samiland refers to Sami traditional territory which extends across the borders of Norway, Sweden, Finland and Russia. Likewise, in northern Africa, the traditional territory of the Tuareg people covers a vast geographical area in the Sahara desert that spans across five nations, from northern Nigeria, to Algeria, Niger, Mali, Burkina Faso, and southern Algeria. The transnational identities of these groups directly correlate with their exclusion from the UNFCCC process where they are sidelined by decision-making processes that are restricted to actors representing nation-states (Brugnach et al., 2017; Roosvall & Tegelberg, 2015; Schroeder & Lovell, 2012). Without formalized political recognition, transnational indigenous actors with observer status can only partake in events "on the side" (Schroeder & Lovell, 2012). This does not offer a sufficient framework for facilitating collaboration and integration of these actors within the UNFCCC process.

The third challenge is *epistemological.* It relates to conflicting ways of knowing about climate change (indigenous versus non-indigenous knowledge) that result, consequently, in conflicting visions of how it should be addressed. Callison (2014, pp. 11–12) differentiates between understandings of climate change in laboratories, policy debates or news coverage and the work that is needed to better "… understand climate change as a problem *in the world.*" Orlove et al. (2014) link this problem to geographical scales, characterizing it as the uneven spatialities of climate change knowledges. Smith and Sharp (2012) argue that while TEK and Western science can be viewed as complementary, the integration of indigenous TEK into climate science and politics has been highly uneven at best. After developing a framework for assessing the degree of inclusion of TEK, they conclude that neither the UNFCCC process nor the Kyoto Protocol are inclusive of indigenous knowledges (Smith & Sharp, 2012). A related study points to a lack of critical engagement with indigenous knowledge systems in the Working Group II portion of the IPCC's 5th Assessment Report (Ford et al., 2016).

Connecting Scales: Indigenous Knowledge in Climate Research and Policy

Global climate models provide a broad picture of climate change and its projected consequences for particular regions and ecosystems. However, they cannot provide an adequate measure of how these changes are experienced locally (Hulme, 2010; Salick & Byg, 2007). Jasanoff sums up the research challenge this poses: "Living creatively with climate change will require re-linking larger scales of scientific representation with smaller scales of social meaning" (2009, p. 238). In response, climate researchers have begun to address this disconnect by bridging global scientific knowledge with local forms of experiential knowledge.

Farbotko and McGregor (2010, p. 162) discuss a division between spaces of rationality and spaces of emotion at UN climate summits. After describing a standard protocol that separates climate science and politics from emotion, they argue that the potential for change increases in rare moments when emotion and affect unsettle this rationalistic framework (2010, p. 164). They conclude that this raises questions concerning the "emotional geographies of climate change" which have been rarely considered in climate change research. Mahony and Hulme call for research frameworks that take into consideration "a 'unique' ecology of social and physical relations" (2016, p. 6). This necessitates examining how space and power intersect at field sites, international conferences, and in computer modeling systems where particular kinds of climate knowledge are produced at the expense of others. Furthermore, they draw attention to openings where "scalar correspondences" could be established by shifting the focus of scientific bodies, like the IPCC, toward the co-production and circulation of more nuanced and relational climate knowledge (2016, pp. 16–17).

In 2007, an unprecedented two-day research symposium on "Indigenous Peoples and Climate Change" took place at the Environmental Change Institute at Oxford University. The event brought together researchers from a range of disciplinary backgrounds to discuss how indigenous peoples are experiencing, observing, interpreting and adapting to climate change. Discussions centred on pathways to greater inclusion of indigenous knowledge and perspectives in climate change policy, research and action. A report on the findings divides the world into six different ecosystems—Polar, Alpine, Deserts, Rainforests, Islands and Temperate—detailing how climate change is experienced locally and how indigenous peoples are "… actively trying to adapt to the changing conditions" (Salick & Byg, 2007, p. 15).

The report concludes by calling for further collaboration between climate scientists, policy-makers and indigenous peoples. A number of techniques are identified that could be implemented to "decolonize methodologies" by including indigenous peoples in each of the stages of research (Salick & Byg, 2007, p. 24; Smith, 1999). Moreover, the authors emphasize that collaborative research and action could ultimately result in enhanced indigenous self-representation in the UNFCCC process, better documentation of indigenous knowledge of climate change (TEK), and the establishment of joint actions and networks between researchers and indigenous peoples (Salick & Byg, 2007, p. 25).

Since then, a growing body of research has sought to bring together scientific and indigenous ways of understanding *local and regional* impacts of climate change and other environmental issues (Bolton et al., 2012; Brugnach et al., 2017; Tauli-Corpuz & de Chavez, 2009; Tipa, 2009). *What we know, don't know, and need to know about climate change in Inuit Nunangat* (Bolton et al., 2012) exemplifies this approach. Community actors are identified as key stakeholders in the research and, in an extensive literature review, it is noted that since 2006 the number of publications that combine scientific and indigenous ways of knowing have increased (Bolton et al., 2012, pp. 26–27). Tipa (2009) demonstrates how Māori knowledge of stream and river dynamics can benefit contemporary resource management if this TEK is effectively communicated and integrated with scientific approaches. Adams (2004) and McGregor (2002) have focused on conservation initiatives that combine indigenous TEK and scientific knowledge in Australia and Canada.

Others contend that at the *global* level, inviting indigenous actors to collaborate by contributing insights to climate policy and knowledge production can help to overcome the mismatching of scales (Brugnach et al., 2017; Termeer & Dewulf, 2014). Brugnach et al. (2017, p. 8) sum this up, relating it to climate mitigation:

> Collaborating across scales is a challenging but crucial ingredient of including indigenous peoples in climate mitigation. This requires arrangements at the governance scale that not only span the local to the global level, but are also interconnected. The interconnected multi-level nature of climate change can be more effectively addressed by creating a better fit between the scale of the problem of climate change and the governance scale at which it is supposed to be addressed … This is another argument for not only looking at local indigenous communities … but also for trying to include their voices more strongly in the global discussion on climate change and ways to address it.

The authors refer to new forms of blended knowledge as an important by-product of finding ways to "collaborate across scales": approaches that could facilitate greater inclusion of indigenous voices in climate governance frameworks *and* in media representations.

In their *Guide on Climate Change & Indigenous Peoples*, Tebtebba Foundation (Indigenous Peoples' International Centre for Policy Research and Education) has developed one such means of collaborating across scales (Tauli-Corpuz & de Chavez, 2009). The guide outlines a policy-oriented approach to bridging scientific and indigenous knowledge with two interrelated aims: (i) to enhance indigenous peoples' knowledge of climate change science and governance so they can advocate for a more active role in policy and decision-making; and (ii) to make non-indigenous peoples more aware of indigenous experiences and insights on climate change (Tauli-Corpuz & de Chavez, 2009, p. 208). One way forward, they contend, is for the UN to adopt an *ecosystem approach* to climate policy "… which acknowledges that decision-making and management of biodiversity are best carried out using the institutions and governance mechanisms most suited at the ecosystem-level, including a recognition of the central role of indigenous peoples" (Tauli-Corpuz & de Chavez, 2009, p. 154). These clarion calls to connect scales through further inclusion, collaboration and co-production between indigenous and non-indigenous ways of knowing inform our subsequent discussion of indigenous geographies of climate change.

Climate Testimonials from the Arctic to the Amazon: Geographies of Climate Change Impacts on Indigenous Peoples

The indigenous activists we interviewed at COP17 and COP21 shared astute observations of the ways in which climate change is currently effecting the regions they inhabit. The map in Figure 3.1 provides an overview of the wide ranging scope of these testimonials; both in terms of similarities and differences in the changes they observe and the diverse geographic regions and ecosystems they represent. In the selection of passages that follow, interviewees observe effects ranging from changes in temperatures, weather patterns and precipitation to seasonal variations, eroding coastlines and rising sea levels to disappearing ice, snow cover, flora and fauna. They consistently describe these effects by linking them to their acute economic, social and cultural impacts. The subsequent discussion is organized by combining testimonials from members of frontline indigenous communities in four distinct geographical regions: Arctic peoples, Island peoples, Forest and Desert peoples.[4] First, we allow the voices of interviewees to largely stand alone, providing only minimal contextual information. Then, we point to some of the recurring themes in testimonials from each geographical cluster. This illuminates complex *human-nature connections*, and shows how climate change impacts simultaneously produce *common and varied experiences* in frontline indigenous communities across the world.

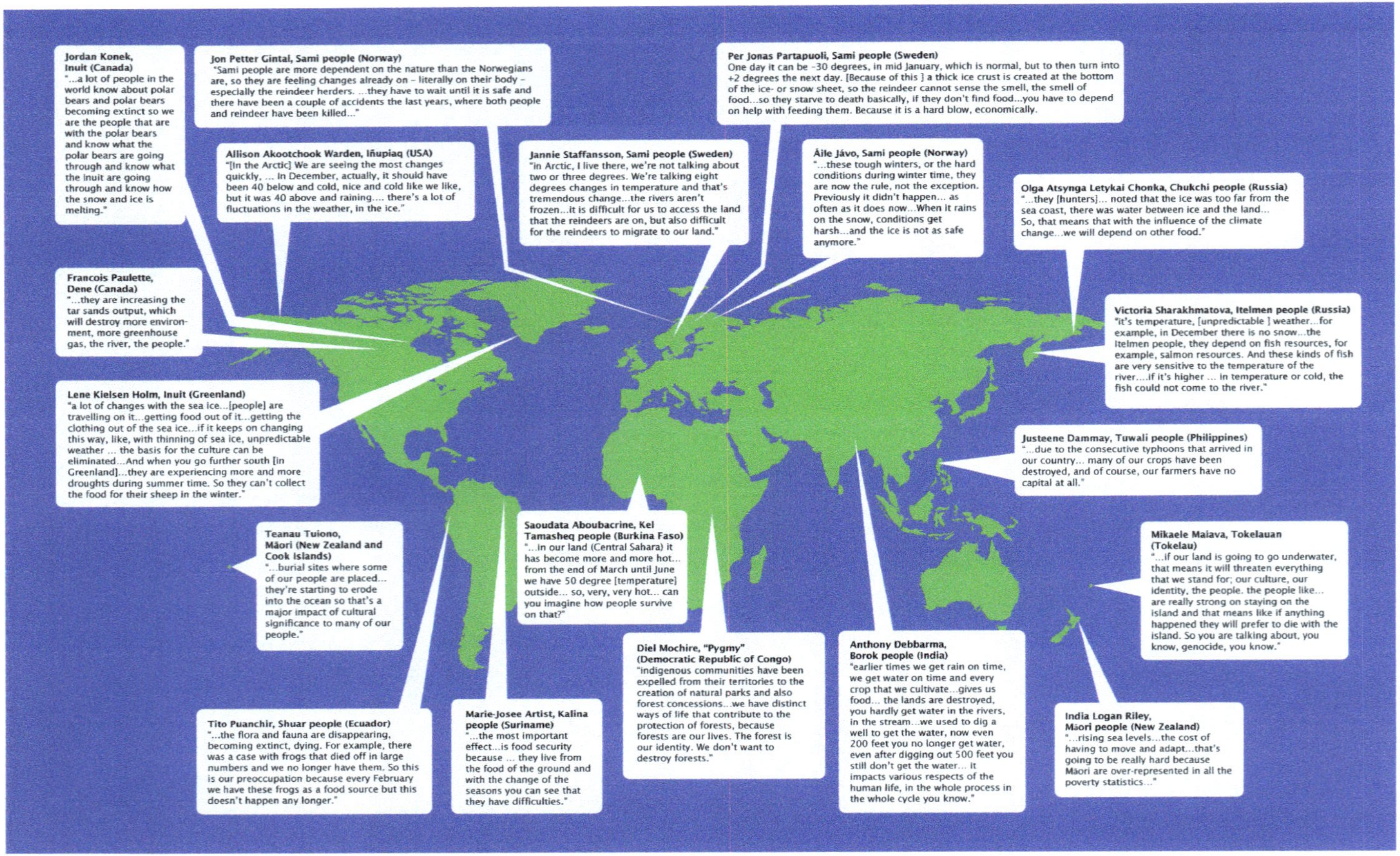

Figure 3.1. Map of indigenous climate testimonials from Arctic to Amazon.

Arctic Peoples

For indigenous spokespersons from the Arctic, dramatic changes in temperature, unpredictable weather and the disappearance of ice and snow were common concerns. As an interviewee from Kaktovik, Alaska put it, "… the Arctic is the most heavily hit by climate change. We are seeing the most changes quickly … In December … it should have been 40 below and cold, nice and cold like we like, but it was 40 above and raining … there are a lot of fluctuations in the weather, in the ice" (Allison Akootchook Warden, Iñupiaq people, USA, COP21). In Kamchatka, Russia, we heard that:

> In December there is no snow or [there are] shifts in the weather, a shift in autumn and winter. For example, my community, the Itelmen people, *they depend on fish resources … salmon resources*. And these kinds of *fish are very sensitive to the temperature of the river*. And, if it's higher in temperature or cold, the fish could not come to the river. (Victoria Sharakhmatova, Itelmen people, Russia, COP21, emphasis added)

Another testimonial, from Western Chukotka, Russia, described the impacts of a changing Arctic climate for local hunters:

> In every village we have some observers of the seal hunting and bear hunting. And, of course, they see big changes in the climate … *they're very connected with hunting, with the seals and sea animals, and these changes are very visible for them* … I live a little bit more in the South, and there was good hunting in my area, but a little bit up to the North, they noticed, noted that the ice was too far from the sea coast, there was water between ice and the land … So, that means that with the influence of the climate change … we will depend on other food. (Olga Atsynga Letykai Csonka, Chukchi people, Russia, COP21, emphasis added)

Two Inuit youth from Arviat, Canada added that "… a lot of people in the world know about polar bears and polar bears becoming extinct, *so we are the people that are with the polar bears* and know what the polar bears are going through and know what the Inuit are going through and know how the snow and ice is melting" (Jordan Konek, Inuit, Canada, COP17, emphasis added). Another interviewee echoed these concerns, stressing that media coverage of Arctic climate change "… tends to focus on, say, the polar bears as opposed to the impact on the people that would be hunting bears or that are dependent on the wildlife on the land." (Tiina Kurvits, Many Strong Voices, Canada, COP17).

The changing Arctic climate is creating new challenges for wildlife and for local people who depend upon these animals for survival. Human-nature connections are made to emphasize these *security risks*. Some spoke of the greater risks hunters and animals are taking in order to find food as familiar prey grow increasingly scarce. Others noted that while the plight of polar bears has become a global icon of climate change, indigenous peoples who "live with the polar bears," and are struggling to adapt to a rapidly changing Arctic ecosystem, must not be ignored.

The social and cultural impacts brought on by rapidly changing temperatures and landscapes were another source of concern. An Inuit spokesperson, from Nuuk, Greenland, noted that:

> Northwest of Greenland … they are seeing a lot of changes with the sea ice. They are both travelling on it … getting food out of it … [and] getting clothing out of the sea ice. So with the changes, you can imagine that *if it keeps on changing this way*, like, *with thinning of sea ice, unpredictable weather … the basis for the culture can be eliminated.* (Lene Kielsen Holm, Inuit, Greenland, COP21, emphasis added)

Others spoke of the risk climate change poses to their livelihood:

> I'm raised as a reindeer herder and hopefully I will be able to work as a reindeer herder in some way in the future too … Since we have this livelihood … climate change is of course substantial, because we live off of nature and the animal. (Per Jonas Partapuoli, Sami, Samiland, Sweden, COP21)

Another Sami interviewee from Sweden reiterated why Per Jonas Partapuoli was able to attend the COP "… because he couldn't go up to the mountains to get the reindeer, to herd the reindeer, because the rivers aren't frozen … Right now it's difficult for us to access the land that the reindeers are on, but also difficult for the reindeers to migrate to our land." She later associated these changes to the land with a range of cultural and psychological impacts:

> The language is based upon nature and also the knowledge from nature and that is changing as well. The identity from handcrafting, the peoples that are doing handcrafting. *It's harder to reach the things that they need from nature* … it is multiple stress, especially for reindeer herders and … children, they are very stressed as well. So it impacts families, communities, distress for climate, *distress for the reindeers getting food.* (Jannie Staffansson, Sami, Samiland, Sweden, COP21, emphasis added)

A Lule Sami delegate from Norway emphasized that "the Sami people are more dependent on nature than the Norwegians are, so they are feeling changes already on, *literally on their body*; especially the reindeer herders." He then offered an example, stressing that

> [they] have to cross some rivers and you need thick enough ice on the rivers to migrate over with the reindeers. So they have to wait until it is safe and there have been a couple of accidents the last years, where *both people and reindeer* have been killed. (Jon Petter Gintal, Lule Sami, Samiland, Norway, COP21, emphasis added)

Reindeer herder and Sami Council President Áile Jávo echoed these sentiments, stressing that:

> When there are huge shifts in temperature, going from over zero degrees to maybe minus 30 and then back to over zero degrees again, it causes … the snow to turn into ice … It becomes very hard for the reindeer to dig for food under the snow … it effects the economy when the Sami must feed the reindeer. (Áile Jávo, Sami, Samiland, Norway, COP21)

Changing weather patterns and disappearing habitats are creating new hazards and stresses for the humans and animals that co-inhabit the Arctic ecosystem. Interviewees describe *economic*, *cultural* and *psychological* stresses ranging from difficulty finding organic materials to make traditional crafts to reindeers exhibiting signs of distress due to difficulty finding sustenance. Others detailed the *corporeal* repercussions of a changing landscape that can result, in the most extreme instances, with the loss of life for humans and reindeer. As one interviewee put it succinctly, if the Arctic climate continues changing at this rate "the basis for the culture can be eliminated" (Lene Kielsen Holm, Inuit, Greenland, COP21).

Island Peoples

Coastal erosion and a corresponding rise in the sea level were of central concern for indigenous representatives from *island ecosystems*. They cited a range of different impacts from culture loss to displacement. One Māori activist, from Hawkes Bay, New Zealand, described consultations with planning experts to determine where to build a new *marae* [community meeting place]. She explained that "… the experts said that it [the land] is not going to be there in 30 years' time, you're going to lose your ancestral land, you're better to build your marae up on the hill. So that is one of the real outcomes." The rising coastline was also linked to disappearing art and culture "… there are a lot of sacred sites, a lot of rock art which is really important, really old pre-contact rock art that will be lost with rising sea levels." Finally, for this interviewee, the most devastating social impact was displacement: "I know my marae is going to have to move somewhere else, they're at or just above sea level … so that is all going to have to move as well with rising sea levels, yeah" (India Logan-Riley, Māori, New Zealand, COP21).

These concerns were echoed by another Pacific islander who mentioned "burial sites where some of our people are placed … [that] are starting to erode into the ocean so that's a major impact of cultural significance to many of our people." He went on to describe industrial developments, such as deep sea oil drilling, "that exacerbate climate change", and concern coastal peoples who "still rely on traditional food sources … out in the ocean" (Teanau Tuiono, Māori, New Zealand and Cook Islands, COP21). A Tokelauan delegate told us that on his island "… the sea level is rising and the salt water actually gets into the soil so that has impacted our crops … it is very hard to grow any, you know any crops there and so that is threatening our food security" (Mikaele Maiava, Tokelauan, Tokelau, COP21).

Another cluster of statements raised concerns about the heightening frequency of super typhoons, powerful cyclones and other extreme weather events. A Tuwali interviewee from Luzon Island in the Philippines and a Pacific islander each convey related concerns:

> So in our community we experience climate change through having strong typhoons. Recently we experienced the strong typhoon Haiyan, which killed many people … due to the consecutive typhoons that arrived in our country … many of our crops have been destroyed, and of course, our farmers have no capital at all … [and] because of the destruction of our lands, our cultural traditions have—are being forgotten by the youth. (Justeene Dammay, Tuwali people, Philippines, COP21)

> I was on the island of Aitutaki visiting some relatives last year and they were talking about how Cyclone Pam had come through and like just destroyed the roof and they were all huddling underneath the ruins of their house and stuff like that. And so what the science tells us is that those cyclones will become more intense over time, and possibly … and that … makes a place difficult to live in, especially if you've got to rebuild your house every five years. (Teanau Tuiono, Māori, New Zealand and Cook Islands, COP21)

These testimonials attest to the devastating social and cultural impacts that rising sea levels and extreme weather are having on island ecosystems. With traditional food sources becoming scarcer, and storms striking with unprecedented magnitude and frequency, the cultural and economic fabric of life in island communities is increasingly at risk. As one Pacific islander captured it uncompromisingly: "You are talking about, you know, genocide … People won't move because that's who they are, that's where they belong. So yes, it's really a huge impact to us and it will destroy everything that we live for" (Mikaele Maiava, Tokelauan, Tokelau, COP21).

Forest and Desert Peoples

Indigenous people who live in *forest ecosystems* provided detailed accounts of the impacts of changing seasons and the disappearance of flora and fauna. Common across these accounts was a shared emphasis on an emerging lack of food security, and the threat this poses to traditional methods of hunting, gathering and subsistence agriculture. An Ecuadorian leader, described the changes indigenous people are witnessing in the Amazon rainforest.

> Flora and fauna are disappearing, becoming extinct, dying. For example, there was a case of frogs that died off in large numbers and we no longer have them. So this is our preoccupation because every February we have these frogs as a food source but this doesn't happen any longer. (Tito Puanchir, Shuar people, Ecuador, COP17)

Similar concerns were described by a representative from Bigi Poika, Suriname:

> [W]hat the communities feel is that the seasons are changing and the most important effect, I think, is more about food security because in Suriname they do a lot of agriculture, so they live from the food of the ground and with the change of the seasons you can see that they have difficulties to decide when to plant, when it is the right time to plant, because sometimes it's too dry and sometimes the rain starts too early. (Marie-Josee Artist, Kalina people, Suriname, COP21)

Others described the general threat posed by the destruction of forests that indigenous groups have inhabited and co-managed for centuries. As an interviewee from the Congo region put it, "We Pygmy, we have been living in forests for thousands of years and we know the forests. So, we have distinct livelihoods that contribute to the protection of forests, because forests are our lives. The forest is our identity. We don't want to destroy forests" (Diel Mochire, Pygmy, Democratic Republic of Congo, COP21). In North-East India and Bangladesh, seasonal variations and industrial developments are effecting the entire cycle of local cultivation:

> In earlier times we got rain on time, we got water on time and every crop that we cultivate in June gave us food. But now, due to deforestation and the building of mega dams … destroying all this biodiversity has actually created an *imbalance of ecology*, ecosystem. Therefore we, the indigenous people, are finding it hard

> to survive because all our forests are taken away, because we don't have a place for the June cultivations. (Anthony Debbarma, Borok peoples, India, COP21, emphasis added)

In these statements the dangers posed by climate change are seen to be exacerbated by harmful human developments, such as logging and damming. A combination of natural effects (changing seasons, unpredictable weather) and industrial developments encroaching upon indigenous lands are making it harder to survive.

In the Sahara desert ecosystem, we heard about extreme temperatures that are having a wholly different range of devastating consequences. A Tuareg interviewee from Burkina Faso summed up these concerns candidly:

> In our land (Central Sahara) it has become more and more hot … from the end of March until June we have 50 degree [temperature] outside … so, very, very hot … can you imagine how people survive in that? […] Every period becomes more difficult … we lose animals, they die … in some areas people die because of no water, and [there are] health issues because of poverty, no food, no milk, no meat. (Saoudata Aboubacrine, Kel Tamasheq people (also known as Tuareg), Burkina Faso, COP21)[5]

Here, once again, human and non-humans are both deeply implicated in the changing natural world. Food shortages, health issues and poverty are all being compounded by a warming climate, in the worst cases the consequence is death.

Traditional Ecological Knowledge

Indigenous peoples representing every ecosystem gave passionate testimonials linking a growing ecological imbalance to the distinct ways in which indigenous peoples relate to nature, particularly through TEK. Some made an explicit connection between climate change, intimate community ties to the land and the loss of TEK.

> It's going to be really hard to hold on to a lot of the cultural areas, the very deep connection with the land, to *lose that land is going to be hard and traumatising and frustrating* because it's not … *it's not Māori's fault at all what's happening*. (India Logan-Riley, Māori, New Zealand, COP21, emphasis added)

> … *when our lands are taken away, everything is gone*. Your culture is gone, your identity is gone, your language is gone. So you are no longer being recognized as a human being. (Anthony Debbarma, Borok peoples, India, COP21, emphasis added)

Others raised concerns about the effects of climate change on the transmission of TEK across generations.

> … *everything goes with climate*, with weather, with seasons, with explanations, you know, so in the long run I am afraid that … I think we need to do extra things, extra activities, *extra to transfer the knowledge of the elders to the children, because otherwise it will be lost*. (Marie-Josee Artist, Kalina people, Suriname, COP21, emphasis added)

> … *we are the generation of climate change*, my generation. We heard stories from our grandfathers and grandmothers of the good winters, and once in a while—like one time in every ten years—they had a bad winter. But, *their definition of a bad winter, that is what we are experiencing all the time*. (Jannie Staffansson, Sami, Samiland, Sweden, COP21, emphasis added)

Others evoked TEK as a means of adapting to changing environmental conditions; often by stressing that indigenous knowledge could play a more significant role in global efforts to mitigate climate change.

> There is a gap between what is implemented at the local level and … the international level, where there are only words in the declaration that are not implemented. So, in my opinion, it is important to consider *traditional knowledge and science* because they go together … that will allow us to recognize indigenous peoples that are not scientists but that can work to promote climate issues. (Diel Mochire, Pygmies of North Kivu, Democratic Republic of Congo, emphasis added)

> *[E]verything is connected*, water is connected with the land, the indigenous peoples preserve their rituals and their cultural links and spiritual links with land … *maybe … in 40 years* the people who discuss here their issues and challenges, *they will come to us and will ask us … how to solve these issues, these problems*. (Olga Atsynga Letykai Csonka, Chukchi people, Russia, COP21, emphasis added)

Human-nature connections are perhaps the most pronounced in these testimonials concerning TEK. TEK is evoked in two ways; firstly, in expressions of the ways TEK is being endangered by climate change; and secondly, in expressions of the ways TEK can and must be embraced as an existing solution to climate-related issues.

While each grouping of testimonials describes distinct local impacts and issues, a common thread across these accounts is a shared emphasis on how climate change is affecting human and non-human actors alike. Natural and social impacts, in other words, are not viewed as inhabiting separate spheres but rather as integral components of an organic whole. This human-nature connection recurred consistently in climate testimonials from the Arctic to the Amazon. Evidently, these testimonials show how patterns of life in frontline indigenous communities intimately linked to the surrounding natural world are being dramatically affected by changing temperatures and landscapes. In the most severe cases, these changes threaten the very survival of the human and non-human inhabitants of these ecosystems. This existential threat is without question what compels indigenous activists and leaders to attend UN climate summits to collectively advocate for climate action. Hence, we shall now turn attention to the geographies of indigenous participation at these events.

Geographies of Indigenous Participation at the COPs

At UN climate summits, the natural and social impacts of climate change are salient features of indigenous narratives. Bjørst (2012) exemplifies how, during these events, a tension between "global and local natures" emerges in climate-crisis narratives that feature indigenous peoples bearing witness to climate change. By typecasting these actors as victims or global spokespersons for the climate, the complex meanings of climate change that play out at the local scale are often misunderstood or ignored (Bjørst, 2012, p. 110; Callison, 2014; see also Roosvall & Tegelberg, 2012, 2013).[6] Hence, despite their powerful testimonials on the natural and social impacts of climate change, representatives of frontline indigenous communities have been largely pushed to the geographical margins of UN climate summits. This section combines participant observation with interview research to examine the geography of indigenous participation at COP17 and COP21. We refer to this as the political geography of climate summits, differentiating between the role of indigenous peoples *inside* and *outside* of the UNFCCC process.

Indigenous Representation Inside the COPs

The structure of the annual UNFCCC climate summits divides participants into two main categories: "Parties" and "Observers". The former refers exclusively to nation-states with a political role in the negotiations that varies according to their expected level of commitment to funding adaptation and mitigation measures. Parties are then divided further into three subcategories:

Annex I Parties (industrialized countries, and countries "in transition"), Annex II Parties (elite OECD countries with specialized financial commitments) and Non-Annex I Parties (developing countries). The categories vary according to the financial commitments these nations are expected to make to the global effort to mitigate the impacts of climate change.

Hundreds of organizations, representing groups with wide ranging interests from the global to the local, are admitted to the COP summits as "Observers". There has been a steady increase in the number of non-nation state actors that attend COP summits, with almost 1,400 on hand at COP15 (Schroeder & Lovell, 2012) and over 2,000 at COP21 in Paris (UNFCCC, 2015). Like "Parties", there are three different subcategories of Observers: i) United Nations System and its Specialized Agencies; ii) intergovernmental organizations (IGOs); and iii) non-governmental organizations (NGOs). With Observer status, these groups are granted access to sessions, side events and exhibits inside the summit (UNFCCC, 2015). However, unlike members of the Parties, they do not have a formal role in the decision-making and are often shut out of high-level meetings and proceedings (Weisser & Muller-Mahn, 2016).

In terms of their recognition within the officially sanctioned UNFCCC process, indigenous peoples—and other transnational groups—either fall within this broad cross section of Observers, or gain access to the proceedings as individual members of a national delegation.[7] Hence, when it comes to formal political representation, indigenous participants occupy a marginal, and sometimes conflicted, position in the deliberations.[8] For those indigenous peoples that do participate in the UNFCCC process, Brugnach et al. (2017, p. 5) note that:

> … the possibilities of having a real impact on the discussions and outcomes appear to be limited—e.g. the inclusion of indigenous knowledge in the UNFCCC and the Kyoto protocol is nearly inexistent … Powerful countries and country coalitions have a much greater impact … Therefore, the representation of indigenous peoples [sic] interests in climate mitigation decision-making depends strongly on the political influence of indigenous peoples on their national governments. Due to historical processes of marginalization, this influence is often very limited.

The authors also describe a "participation paradox" (Brugnach et al., 2017, p. 11) where indigenous representatives—who spend considerable time away from their constituencies advocating on their behalf—can themselves become distanced from the people and issues they represent. In *The Right to Be Cold*

(2015), Canadian Inuit activist and Noble Prize nominee Sheila Watt-Cloutier documents how she has grappled with this paradoxical situation in her ongoing struggle for Arctic climate justice.

Of the 17 indigenous activists we interviewed in Paris (2015), 11 indicated that they carried the status of "Observers" or "Party" delegates with at least some degree of access to deliberations *inside* the summit (see Chapter 1, Table 1.2). However, several commented on the conflicted nature of their position as delegates, or on the limited access they had to meetings where decisions were being made with consequences for indigenous peoples.

At COP17, Sami parliament member Vibeke Larsen commented on the challenge she faced balancing the interests of her constituents with the Norwegian government, since she was included in the official Norwegian delegation. In discussions concerning renewable energy initiatives, she recounted having difficulty reconciling Norway's position with local Sami interests. Developments such as windmills, Larsen explained "… come into a conflict with Sami reindeer herding … between climate and the traditional way to live. So that's where the Sami parliament comes in … we consult with the Norwegian society, on how can we do this without [having] too much influence on the reindeer herding but still have the green energy." This, she later stressed, is "… a hard place for any [member of the] Sami parliament because we want to be responsible for society, we want to be responsible for our future and, at the same time, we have to take care of our old culture and the reindeer herding, yes" (Vibeke Larsen, Sami, Samiland, Norway, COP17).

This account illuminates a dichotomy between *local* interests (of Sami reindeer herders), and Norway's *national* commitment to curbing *global* greenhouse gas emissions through the expansion of renewable energy infrastructure. The participation paradox, in other words, lies in Larsen's conflicted hopes of simultaneously being responsible to the broader society without harming a local cultural practice.

Similar conflicts emerged for indigenous activists we spoke with at COP21. Lene Kielsen Holm, a delegate with the Inuit Circumpolar Council (ICC), described having access to some but not all of the summit halls. For instance, her status as an observer with a transnational political organization, restricted her access to events strictly sanctioned for national delegates with "Party" status, in spite of their relevance to issues of critical significance to her advocacy work for the ICC.

At COP21, Jannie Staffansson described a correlation between the transnational governance structure of the Sami Council and the challenges she faced getting access to the official meetings where binding agreements are produced at the COPs. To exemplify this she recounted two conversations with UNFCCC Executive Secretary Christiana Figueres where Staffansson had tried, unsuccessfully, to convince Figueres of the importance of including indigenous peoples in the negotiation process. Staffansson then contrasted this lack of access with a precedent set for inclusion at Arctic Council meetings where indigenous peoples have a say in the drafting of legally-binding agreements. "That is why this work [at UN summits] is also frustrating because you don't have the right to speak up, you don't have the right to be a part of the negotiations and what they are drafting and whatever the outcome is … but in the Arctic Council you do and I am quite used to that" (Jannie Staffansson, Sami, Samiland, Sweden, COP21). At COPs, the negotiating environment is thus a source of geographical injustice since transnational indigenous actors are excluded from the process because their identities do not conform to the established governance structure.

Marie-Josee Artist, a indigenous activist from Suriname who has been attending COPs since 2008, reflected on attending COP21 without accreditation:

> Before, I did some advocacy and I was in the Indigenous Peoples' Caucus … I kind of miss it now because I don't know what is going on in the big area … I didn't think that it was a big issue, but now that I am not accredited through another organization it feels like a loss, so yes I would have been there. (Marie-Josee Artist, Kalina people, Suriname, COP21)

The International Indigenous Peoples Forum on Climate Change (IIPFCC) was established in 2008, shortly after the adoption of United Nations Declaration on the Rights of Indigenous Peoples (2007). At COP summits, the IIPFCC takes the form of a Caucus dedicated to representing indigenous peoples who participate in the UNFCCC process. The caucus is divided into seven regions—Arctic, Russia and Eastern Europe, Africa, Asia, Pacific, North America and Latin America and Caribbean. There are also local steering committees composed of members from within each of these regions. On the website, the IIPFCC outlines its mandate:

> … to come into agreement specifically on what IPs [indigenous peoples] will be negotiating for in specific UNFCCC processes. IP representatives attending the meetings have their own organizations at subnational, national and global levels which have their own agenda, priorities and own proposals that they may carry and push for during the IP Caucus meetings. (IIPFCC, 2017)

Six interviewees described work for the IIPFCC Caucus, ranging from morning briefings to communication and the planning of demonstrations (outside the summit), as an integral part of their daily routine at COP21. According to one interviewee, this work entailed "a lot of lobbying", statement writing and interventions. She also detailed the collaborations that take place between Caucus members with varying degrees of access to the summit halls:

> Some of the delegates here, especially from the Pacific Islands, they have gotten slots for being party members as well as party observers. So, they are able to attend the sessions and then they can report to us … but they are not able to negotiate or to have a say in the negotiation … That's a good way to do it right now, but in the very end they're not representative, they're now under the states and that's not self-governance, that's not long term, but it's a last way of us getting to know what's happening in the room, it's a last solution … our plan F. (Jannie Staffansson, Sami, Samiland, Sweden, COP21)

Another caucus member described the task of working to ensure the "concerns that indigenous peoples have on the *inside*" lined up with others who "will be *outside* working with civil society in more traditional activist roles." He emphasized that "… working together to make sure that our messages are consistent but also as visible and as clearly heard as possible …" was critical (Teanau Tuiono, Māori, New Zealand and Cook Islands, COP21). Another member stressed that, in spite of the regional divisions, members of the IIPFCC Caucus worked "in solidarity with each other because we are indigenous peoples and, to a great extent, we are in the same situation" (Lene Kielsen Holm, Inuit, Greenland, COP21). The situation these actors confront illuminates the paradox of indigenous participation, demonstrating how the summit geography has the effect of setting "… limits on what can be spoken; the social space conditions what is heard" (Meusburger, Livingstone, & Jöns, 2010, p. 75).

This geography of injustice pervaded the summit spaces in our participant observation of indigenous activism at COP17 and COP21. At both COPs, we found indigenous activists relegated to the geographic margins in terms of their proximity to formal decision-making processes, decision-makers, and to the mainstream media outlets covering the official deliberations.

Outside of the officially sanctioned meeting halls, we observed and conducted interviews with several indigenous activists who either had limited access (Observer status) or no accreditation at all. These actors cited various financial means of getting to the COPs, including institutional support from academic research grants, NGOs and philanthropy foundations, to online crowd funding campaigns. With restricted access, side events in spaces outside the confines of the summit halls became *the* spaces where these activists participated most actively during the COPs.

At COP17, the indigenous organizations and activists we encountered resided on the margins of the conference centre in a large poster room. Several booths representing indigenous peoples and their concerns about climate change were randomly dispersed throughout this large conference room, with no common space designated to showcase indigenous issues. At press conferences inside the summit, planned by the Inuit Youth Delegation, Tebtebba and the Indigenous Environmental Network, we sat in a room that was at times virtually empty. In one instance, after giving detailed presentations connecting climate impacts to human rights and justice, indigenous activists representing four different regions were met with silence. The few reporters that were on hand asked no questions, and quickly rushed away to get to the next events.

At COP21, indigenous peoples had a stronger collective voice and presence, however their marginalization within the structure of the summit was by no means overcome. Located in the Climate Generations Space, the Indigenous Peoples' Pavilion was a considerable distance from the "Blue Zone" where the official meetings took place. Weisser and Muller-Mahn refer to the location of the IIPFCC pavilion to exemplify the uneven "micro-geography" which they argue characterized the Paris summit:

> It is at least likely that marginal voices have been affected adversely. For example, the pavilion of the International Indigenous Peoples' Forum on Climate Change (IIPFCC) was only granted a space in the adjacent Climate Generations Area, where a very limited number of decision-makers passed by. The private sector, on the other hand, was provided with space for delegation offices and pavilions within the official COP premises, involving companies like Facebook, or associations like the International Air Transport Association. (2016, p. 12)

This juxtaposition demonstrates the extent to which indigenous peoples were marginalized from the spaces of deliberation where decision-makers and corporate stakeholders worked to hash out the Paris Agreement.

Throughout the week, at the Indigenous Peoples' Pavilion we attended regional days with programs dedicated to showcasing indigenous cultures and raising awareness of climate-related issues in local communities. On the regional day for North America, for example, indigenous activists and leaders spoke on a wide range of topics including "Traditional Knowledge and Climate

Change"; "Climate Change and the Inherent and Treaty Rights of Indigenous Peoples" as well as a film screening and roundtable discussion titled "North American Indigenous Peoples on the Frontlines of the Climate Crisis". During these events, we witnessed symbolic gestures made by national and international leaders, including French President Francois Hollande; Tuvalu's Prime Minister, Enele Sopoaga, Mary Robinson, Special Envoy of the UN Secretary General on Climate Change; and Canada's Minister of Environment Catherine McKenna. These "high-profile visitors" attended cultural presentations, spoke informally with indigenous leaders, or gave brief speeches on regional days related to their national interests. In these moments, an emphasis on goodwill and the celebration of indigenous culture, in a space already on the margins of the summit, tended to overshadow the more concrete demands of indigenous climate justice activists.

This symbolic recognition was perhaps best exemplified shortly after the indigenous pavilion opened on December 1st. As we stood waiting between interviews, the mood in the pavilion abruptly changed from casual to high energy as French President Francois Hollande made his way into the space, accompanied by an entourage of security and press. His arrival, seemingly unannounced, created a great deal of excitement. After entering the pavilion, Hollande spent a few minutes conversing and shaking hands with indigenous leaders before being quickly rushed away by security personnel. Áile Jávo, President of the Sami Council, spoke, by contrast, about indigenous peoples only being mentioned by two of 135 nation-state representatives in the introductory speeches on the first day. These were the leaders from Canada and Tuvalu (Áile Jávo, Sami, Samiland, Norway, COP21). Another symbolic gesture we encountered inside the summit was indigenous leaders dressed in traditional clothing posing together for photographs while facing a media scrum. This symbolic treatment, as we have argued elsewhere (Roosvall & Tegelberg, 2012, 2015), emphasizes the recognition of identity while displacing the recognition of status (Fraser, 2000, 2008), particularly the right to negotiate on a par with other nations in the UNFCCC process.

While symbolic acts like those described above permeated the space in and around the Indigenous Peoples' Pavilion, any actions or displays that exceeded what was deemed appropriate were quickly supressed. Figure 3.2 depicts a large banner that reads "Indigenous Peoples Defending Mother Earth." The banner was erected on one of the summit walls by a group of indigenous activists. As people began gathering around it, security personnel quickly moved in to remove the banner and disperse the crowd. This demonstrates the crucial connection between space and power that governs what can be heard and said within the walls of the summit (Foucault, 2010; Meusburger et al., 2010).

Figure 3.2. Indigenous activists surround banner inside COP21.

This geographical marginalization at climate summits demonstrates that, in spite of the UN Declaration on the Rights of Indigenous Peoples (UNDRIP), there is not yet full or effective inclusion of indigenous peoples in UNFCCC decision-making processes. The distancing effect inside summits is compounded by the geographical remoteness, in a physical sense, of many frontline indigenous communities. Although they may already be confronting the direct consequences of climate change, transnational indigenous groups have yet to be granted access on par with nation-states inside the COP process. However, this does not mean that indigenous peoples were not active elsewhere making demands for climate justice. It is to mapping and describing some of these manifestations that our attention shall now turn.

Indigenous Representation Outside the COPs

At COP summits, the most contentious politics have taken shape outside the conference centres in a "variety of autonomous self-managed 'free' spaces" (Chatterton et al., 2012, p. 603) across the city where activists meet, eat, sleep and plan their actions. Chatterton et al. argue that at COP15, in December 2009, it was during diverse mobilisations around the city of Copenhagen that climate justice emerged as an antagonistic discursive framing (Chatterton et al., 2012; see also Ytterstad & Russell, 2012). This contrasted with the dominant "post-political" consensus logic[9] that structured debates inside the summit. It was in spaces outside the Copenhagen summit, in other words, where climate justice entered into activist vocabulary and began to politicize the terms of climate change debates (Chatterton et al., 2012).

During our field research at both COPs, we found that the demands of indigenous activists resounded much louder in activist spaces and demonstrations *outside* the summit halls. In Durban, the University of KwaZulu-Natal hosted an alternative climate summit and designated its campus the "COP17 People's Space". This parallel conference took place in conjunction with the official summit, featuring daily keynote speakers, book launches and panel discussions that placed emphasis on themes with a climate justice orientation that were not prominent inside the COP. On December 3, thousands of people marched through the streets of Durban calling for climate justice. Tom Goldtooth, executive director of the Indigenous Environmental Network, was among the indigenous activists who attended the march. When a *Democracy Now!* reporter asked why he was taking part in the march, Goldtooth responded that "We're here to liberate Mother Earth, to liberate the trees, to liberate the soil in the ground, to stand together with the farmers … to stand together with the people of the world, demanding real change, systemic change, not climate change" (Democracy Now!, 2011a). On December 6, members of the Indigenous Environmental Network took part in a protest at the entrance to the summit where they displayed a Canadian flag covered in oil to represent the devastating environmental impacts of the Alberta Tar Sands. They were interviewed and photographed by media organizations during the demonstration.

In Paris, the programs at general spaces of activism, such as Place to Be and Climate Action Zone, included climate justice themed events that sometimes featured indigenous activists and leaders. For example, on December 1–2 the Guardians of Mother Nature Alliance Summit organized round table discussions on topics such as "Recognition of Ancestral Territorial Rights: A Solution for the Protection of the Biosphere"; "Destruction of the Ecosystem: Moving Toward Climate Chaos"; and "False Solutions that Destroy Indigenous Peoples and Their Environment".

At the International Rights of Nature Tribunal on December 4–5, citizens from around the world were invited to testify publicly to the destruction of the Earth. The aims of the event, organized by several large NGOs (Global

Alliance for the Rights of Nature, End Ecocide on Earth, NatureRights, and ATTAC France), were to enable "… indigenous and local communities to share their unique concerns and solutions about land, water, culture with the global community" and to "… provide a vehicle for reframing and adjudicating environmental and social justice cases within the context of a Rights of Nature based earth jurisprudence" (International Rights of Nature Tribunal, 2017).

In public spaces around the city, networks of indigenous activists organized their own press events and public demonstrations.[10] On December 1, we attended a Forest Day Press Event where indigenous leaders from forest ecosystems in Indonesia, the Amazon, Central America and the Democratic Republic of Congo gave a joint press briefing. The leaders joined forces to launch a global research map that quantifies the large volume of carbon stored on indigenous land in tropical rainforests. A press release, written to coincide with other Forest Day events at COP21, emphasized that:

> Research is beginning to show that forest peoples *armed with strong rights are able to outperform any other forest managers*—public or private. Indigenous peoples and local communities worldwide own 50 percent of the global land mass, according to a recent study, yet they control less than 10 percent of the land they claim—suggesting a dramatic failure to take advantage of an existing solution to climate change. (Woods Hole, 2015, emphasis added)

Presentations drew on data from a research collaboration between scientists, policy experts and "forest guardians" to advocate for national leaders to recognize community forest rights and consider these indigenous communities an "existing solution" to curbing climate change. These efforts resemble a distinction Naomi Klein (2014, p. 375) has made elsewhere concerning the legal power indigenous communities have to protect their lands from harmful development and the obstacles they face having these rights recognized. In a Q&A session after the presentation, several panelists emphasized the importance of collaborations between scientists and indigenous peoples, stressing that these joint efforts can serve to heighten public awareness of indigenous issues and enhance understandings of climate change effects. It was concluded that if indigenous peoples collaborated with science then perhaps the rest of the world would finally listen.

On December 6, the Indigenous Environmental Network and Amazon Watch organized an "indigenous flotilla for climate justice." The action featured a coalition of indigenous peoples of the Americas who crafted traditional water vessels in their communities and transported them to Paris. Dozens of canoes and kayaks travelled along a canal through the heart of

Paris creating a powerful spectacle and drawing the attention of international media (*Democracy Now!*—United States, *Guardian*—United Kingdom and *National Observer*—Canada, among others). The core message of the demonstration was to demand legally binding recognition of indigenous rights in the final language of the Paris Agreement (Democracy Now, 2011b).[11] Clearly, it was at events and demonstrations outside the officially sanctioned summit halls where difficult questions emerged concerning indigenous rights and their collective demands for climate justice.

Conclusion

> [C]limate change arguably displaces the very notion of community by displacing human beings, both as a species and as a source of norms, in favor of an impersonal, but naturalized, object of concern … The question today is how to replace the 'not-but' dichotomy of systemic versus specific framings with the integrative logic of 'both-and': how can climate be seen both as a matter of situated 'human activity and edifice' and as a function of greenhouse gases mingling with the 'clouds, oceans, greenery, and soils' that pattern the Earth's inanimate face? (Jasanoff, 2009, pp. 237–238)

Writing on the problem of scale climate change poses, Sheila Jasanoff echoes Michael Maniates concerns regarding the individualization of environmental consciousness introduced at the outset of this chapter. The challenge for both these authors is getting from the individualization (consumer choices) and abstraction (not-but) of environmental problems toward notions of collective responsibility that resonate with broader cross-sections of people. In this chapter, connecting the scales of indigenous climate knowledge with international climate governance structures points to one of the pathways toward enhancing collective knowledge of and responses to climate change. However, it is also clear that much work remains before such "scalar correspondences" (Mahony & Hulme, 2016) can be widely adopted in a global "consciousness of sustainability"—as Lene Kielsen Holm names it at the beginning of this chapter.

Geopolitical efforts to grapple with pressing global crises, such as climate change, increasingly transnational and multi-scalar in scope, have largely remained fixated on dominant national and international spatial categories. This scalar divergence is thrown into sharp relief by questions of indigenous participation as transnational subjects in internationally sanctioned climate governance processes. Evidently, the experiences of climate change in frontline indigenous communities are a source of global injustice in relation to Fraser's (2008, 2014) outline of the economic, cultural and political aspects of injustice. The indigenous activists we interviewed bear witness to each of these aspects: economic loss (e.g. having to feed reindeer when they cannot feed themselves), cultural loss (e.g. burial grounds and vast territories of cultural meaning disappearing into the ocean), and political deficiency (e.g. the exclusionary geopolitics of the summits as space). Thus redistribution (economic remedy), recognition (cultural remedy) and representation (political remedy) are all needed to amend the climate injustices experienced by interviewees. Measures must be taken toward the inclusion of indigenous peoples, and other transnational subjects, in media, politics and research in order to develop better frameworks for understanding the tension between different extensions of space, their relationships to people and to questions of justice. This would enable transnational subjects to advocate for a *redistribution* of the means necessary to adapt to climate change effects; and moreover facilitate the *recognition* of their rights, knowledge and experience; and, lastly, grant them access to just *representation* in the appropriate policy-making and political arenas, as well as in the media.

Climate testimonials attest to the strong human-nature connection central to the transnational identities of indigenous activists. Moreover, these human-nature connections are expressed in ways that are simultaneously spiritual and political. This was evidenced by the strong connections made between TEK and the urgent need for climate action, particularly in instances where indigenous activists representing diverse geographical regions identified themselves as guardians or protectors of the surrounding natural world. These collective notions of guardianship bridge local and global scales. For indigenous activists, in other words, it is imperative to maintain a balanced and sustainable relationship with nature at all scales from environmental stewardship at the local level to global efforts to raise consciousness (in politics and the media) of the threat climate change poses to all of humanity.

The political geography of the COPs showcased the spatial marginalization of indigenous activists inside the summits. Yet, in spite of this formal sidelining, transnational networks of indigenous activists carved out spaces to articulate narratives that oppose the instrumental rhetoric and formalized structures that govern UN climate summitry. It is crucial to bring these powerful voices and narratives from *outside to inside* by recognizing indigenous rights and including indigenous voices within the UNFCCC process. In the next two chapters, we elaborate on these challenges by examining media and the efforts indigenous activists are making to overcome such barriers to inclusion.

Notes

1. For more on collective responsibility, see Chapters 2 and 6.
2. For further analysis of the moral and geographical injustices that arise from the uneven geographical scales identified in this chapter, see Chapter 5.
3. We characterize frontline indigenous communities as those situated in close proximity to current climate change effects that threaten the livelihood, and in some cases the very survival, of these communities. These include rising sea levels, melting ices, droughts, floods and other acute climate change effects.
4. These distinctions are not completely exclusive as some indigenous groups, or subgroups of certain indigenous peoples, may relate in diverse ways to several ecosystems. Sami people, for instance, are recognized as Arctic people through their inclusion in the Arctic Council, among other things. However, some Sami also relate closely to forested areas.
5. In an e-mail message in April 2017, Aboubacrine asked us to modify the quote slightly in order to provide further clarification. This modified version is published here.
6. For further discussion of witnessing, see Chapter 5.
7. For example, in Durban we interviewed Vibeke Larson, at that time a member of the Sami Parliament in Norway (she subsequently became its President), who attended the COP summit with the Norwegian delegation. Likewise, in Paris we interviewed several indigenous representatives who carried "Party" status as part of a national delegation (see Table 1.2).
8. For further examination of this conflicted position, see Chapter 5.
9. See Swyngedouw (2007, 2010) for a detailed account of this logic and its pertinence to UN climate summits.
10. In Chapter 1, we described the strong presence of indigenous activists at a human chain manifestation that took place in the streets of Paris on November 29. We also recounted an indigenous healing ceremony outside the Bataclan theatre that linked the experiences of indigenous communities on the frontlines of climate change to the devastating loss experienced by victims of the Paris terror attacks.
11. For further discussion of this demonstration and indigenous political activism at the COPs, see Chapter 5.

References

Adams, M. (2004). Negotiating nature: Collaboration and conflict between aboriginal and conservation interests in New South Wales, Australia. *Australian Journal of Environmental Education*, *20*(1), 3–12.

Bjørst, L. R. (2012). Climate testimonies and climate crisis narratives: Inuit delegated to speak on behalf of the climate. *Acta Borealia*, *29*(1), 98–113.

Bolton, K., Lougheed, M., Ford, J., Nickels, S., Grable, C., & Shirley, J. (2012). *What we know, don't know, and need to know about climate change in Inuit Nunangat: A systematic literature review and gap analysis of the Canadian Arctic*. Ottawa: Inuit Tapiriit Kanatami.

Brugnach, M., Craps, M., & Dewulf, A. (2017). Including indigenous peoples in climate change mitigation: Addressing issues of scale, knowledge and power. *Climatic Change*, *140*(1), 1–14.

Callison, C. (2014). *How climate change comes to matter: The communal life of facts*. Durham, NC: Duke University Press.

Chatterton, P., Featherstone, D., & Routledge, P. (2013). Articulating climate justice in Copenhagen: Antagonism, the commons, and solidarity. *Antipode*, 45(3), 602–620.

Democracy Now! (2011a, December 5). Thousands march at U.N. climate summit in Durban to demand climate justice. *Democracy Now!* Retrieved from https://www.democracynow.org/2011/12/5/thousands_march_at_un_climate_summit

Democracy Now! (2011b, December 7). We Do Not Want to Die in Silence: Indigenous People Demand Rights as Draft Climate Deal Reached. *Democracy Now!* Retrieved from https://www.democracynow.org/2015/12/7/we_do_not_want_to_die

Farbotko, C., & McGregor, H. V. (2010). Copenhagen, climate science and the emotional geographies of climate change. *Australian Geographer*, *41*(2), 159–166.

Ford, J., Cameron, L., Rubis, J., Maillet, M., Nakashima, D. Willox, A., & Pearce, T. (2016). Including indigenous knowledge and experience in the IPCC assessment reports. *Nature Climate Change*, 6, 349–353.

Foucault, M. (2010). *The government of self and others: Lectures at the Collège de France 1982–1983*. London: Palgrave Macmillan.

Fraser, N. (2000). Rethinking recognition: Overcoming displacement and reification in cultural politics. *New Left Review*, 3, 107–120.

Fraser, N. (2008). *Scales of justice: Reimagining political space in a globalizing world*. New York, NY: Columbia University Press.

Fraser, N. (2014). Transnationalizing the public sphere: On the legitimacy and efficiency of public opinion in a post-Westphalian world. In K. Nash (Ed.), *Transnationalizing the public sphere* (pp. 8–42). Cambridge: Polity.

Hulme, M. (2010). Problems with making and governing global kinds of knowledge. *Global Environmental Change*, 20(4): 558–564.

International Indigenous Peoples' Forum on Climate Change. (2017). *About the International Indigenous Peoples' Forum on Climate Change*. Retrieved from http://www.iipfcc.org/who-are-we/

International Rights of Nature Tribunal. (2017). *Global alliance for the rights of nature*. Retrieved from http://therightsofnature.org/rights-of-nature-tribunal/

Jasanoff, S. (2009). A new climate for society. *Theory, Culture & Society*, *27*(2–3), 233–253.

Klein, N. (2014). *This changes everything: Capitalism vs. the climate*. New York, NY: Simon & Schuster.

Mahony, M., & Hulme, M. (2016). Epistemic geographies of climate change: Science, space and politics. *Progress in Human Geography*, 1–30. doi:10.1177/0309132516681485

Maniates, M. (2002). Individualization: Plant a tree, buy a bike, save the world? In T. Princen, M. Maniates, & K. Conca (Eds.), *Confronting consumption* (pp. 43–66). Cambridge: MIT Press.

Mann, M. E. (2012). *The hockey stick and the climate wars: Dispatches from the front lines*. New York, NY: Columbia University Press.

McGregor, D. (2002). Traditional ecological knowledge and the Two-Row Wampum. *Biodiversity*, 3(3), 8–9.

Meusburger, P., Livingstone, D., & Jöns, H. (2010). *Geographies of science*. Amsterdam: Springer.

Moser, S. C. (2010). Communicating climate change: History, challenges, process and future directions. *WIREs Climate Change*, *1*(1), 31–53.

Orlove, B., Lazrus, H., Hovelsrud, G. K., & Giannini, A. (2014). Recognitions and responsibilities: On the origins and consequences of the uneven attention to climate change around the world. *Current Anthropology*, 55(3), 249–275.

Pearce, F. (2010). *The climate files: The battle for the truth about global warming*. London: Guardian Books.

Roosvall, A., & Tegelberg, M. (2012). Misframing the messenger: Scales of justice, traditional ecological knowledge and media coverage of indigenous peoples and climate change. In E. Eide & R. Kunelius (Eds.), *Media meets climate: The global challenge for journalism* (pp. 297–312). Gothenburg: Nordicom.

Roosvall, A., & Tegelberg, M. (2013). Framing climate change and indigenous peoples: Intermediaries of urgency, spirituality and de-nationalization. *International Communication Gazette*, *75*(4), 392–409.

Roosvall, A., & Tegelberg, M. (2015). Media and the geographies of climate justice: Indigenous peoples, nature and the geopolitics of climate change. *TripleC: Communication, Capitalism & Critique*, *13*(1), 39–54.

Salick, J., & Byg, A. (2007). *Indigenous peoples and climate change*. Oxford: Tyndall Centre for Climate Change Research.

Schroeder, H., & Lovell, H. (2012). The role of non-nation-state actors and side events in the international climate negotiations. *Climate Policy*, *12*(1), 23–37.

Smith, L. T. (1999). *Decolonizing methodologies: Research and indigenous peoples*. London: Zed Books.

Smith, H. A., & Sharp, K. (2012). Indigenous climate knowledges. *WIREs Climate Change*, 3(5), 467–476.

Swyngedouw, E. (2007). Impossible "sustainability" and the post-political condition. In D. Gibbs, & R. Krueger (Eds.). *The sustainable development paradox: Urban political economy in the United States and Europe* (pp. 13–40). New York, NY: Guilford Press.

Swyngedouw, E. (2010). Apocalypse forever? Post-political populism and the spectre of climate change. *Theory, Culture and Society*, *27*(2–3), 213–232.

Tauli-Corpuz, V., & de Chavez, R. (Eds.). (2009). *Guide on climate change and indigenous peoples*. Baguio City: Tebtebba Foundation.

Termeer, C., & Dewulf, A. (2014). Governance of wicked climate adaptation problems. In J. Knieling, & W. Filho (Eds.). *Climate change governance* (pp. 27–39). Berlin: Springer.

Tipa, G. (2009). Exploring indigenous understandings of river dynamics and river flows: A case from New Zealand. *Environmental Communication: A Journal of Nature and Culture*, 3(1), 95–120.

UNFCCC. (2015). *Parties & Observers*. Bonn, Germany. Retrieved from http://unfccc.int/parties_and_observers/items/2704.php

Watt-Cloutier, S. (2015). *The right to be cold: One woman's story of protecting her culture, the Arctic, and the whole planet*. Toronto: Penguin Random House.

Weisser, F., & Muller-Mahn, D. (2016). No place for the political: Micro-geographies of the Paris Climate Conference 2015. *Antipode*, 1–19. doi:10.1111/anti.12290

Woods Hole Research Centre. (2015). *Leaders present research quantifying the carbon stored on indigenous territories in the world's largest tropical rainforests on forest day; Release of statement from Indigenous Leaders to Heads of Government* [Press release]. Retrieved from http://www.burness.com/wp-content/uploads/2015/11/Woodshole-media-advisory-ENGLISH.pdf

Ytterstad, A., & Russell, A. (2012). Pessimism of the intellect and optimism of the will: A Gramscian analysis of climate justice in summit coverage. In E. Eide, & R. Kunelius (Eds.), *Media meets climate: The global challenge for journalism* (pp. 247–262). Gothenburg: Nordicom.

· 4 ·

SUMMIT JOURNALISM, INDIGENOUS PEOPLES AND DIGITALIZATION

A Media Ecology Perspective

> I love the media. I make friends with them as much as I can because … indigenous communities are usually marginalized, so it is important to get indigenous voices right at the front, and the media can help do that …
>
> I don't necessarily trust the news that is coming from the mainstream media … they suppress events, or don't cover them, or there are things … that happen on Twitter, but … [you] never see them in the mainstream media.
>
> —Allison Akootchook Warden, Iñupiaq people (USA, COP21)

At UN climate conferences, indigenous peoples from around the world travel vast distances in the hopes of having their voices heard, their experiences and knowledge recognized, and their demands for climate justice answered. Hence when asked for their thoughts on media, and more specifically, on its role in communicating about climate change and its local effects, the indigenous activists we interviewed at COP17 (Durban, 2011) and COP21 (Paris, 2015) had much to say. Warden's statements capture the conflicted feelings many interviewees expressed about the media's efficacy in generating awareness of the climate challenges they face at home. On one hand, she emphasizes the importance of *making connections* with mainstream media, since doing so can help foreground indigenous voices and issues that are otherwise marginalized. On the other hand, Warden identifies a critical *disconnection*, describing what makes mainstream media untrustworthy and identifying with social media platforms, like Twitter, that help showcase these limitations, and thereby function as new channels for indigenous expression.

This chapter takes a closer look at the place of indigenous activists, and indigenous peoples more generally, within the news media ecology that surrounds UN climate summits. We place particular emphasis on where indigenous peoples are rendered *invisible*, made *visible*, or make themselves visible in three different parts of the media ecology (see below). To do so, we ask: How do indigenous activists grapple with a paradoxical situation they face in relation to media at UN COPs? How can that situation be characterized more particularly in terms of (in)visibility in diverse types of content? And how do the answers to these questions relate to media responsibility?

Theoretically, the chapter contributes to ongoing discussions of contemporary global power relations; particularly in terms of the potential for transnational groups to make their voices heard and take part in public discourse on a par with everyone else; i.e. that their status is recognized and that representation is just (Fraser, 2008). It also builds on and develops media ecology theories (e.g. McLuhan, 1964; McLuhan, & McLuhan, 1988; Postman, 2000; Scolari, 2013; Tracy, 2010, 2012). Media ecology theorists have brought the science of ecology into the domain of culture and communication (Postman, 2000, p. 10). The metaphor enables scholars to study relations between human beings and mediated environments (i.e. languages, numbers, images, machines), placing emphasis on what gives "a culture its character and, one might say, helps a culture to maintain symbolic balance" (Postman, 2000, p. 11). In other words, media ecology is dedicated to studying the consequences of relationships that develop between humans and the environments in which they communicate.

The ecology of news surrounding the COP summits consists of diverse media forms and actors. Building on our previous research as well as interviews conducted in the field, here we limit our focus to three parts of this media ecology: *print news media*, *social and connective media*, more specifically networked communication on Twitter; and *indigenous produced media* crafted for each of the other two parts. The chapter is structured according to three main themes that permeated much of this material: the centrality and limitations of legacy media; the rise of social media as a news source;

and the increasing use of digital media tools for indigenous self-representation. We begin by considering some of the relevant literature on indigenous media production, on the evolving role of media in environmental protest, and on the growing significance of digital media for indigenous activism.

Confronting Invisibility: The Rise of Indigenous Media

In broader struggles for justice, indigenous peoples have long faced challenges getting their voices heard in mainstream news media. Studies of traditional media in several different national contexts have documented this tendency and demonstrated how problematic colonial stereotypes continue to shape public perceptions of indigenous issues (Anderson & Robertson, 2011; Meadows & Avison, 2000; Pietikäinen, 2003; Pratt, 2008). Anderson and Robertson (2011) systematically examine and critique a hegemonic colonial ideology that has dominated Canadian newspaper coverage of Aboriginal peoples since 1869. In spite of their deep implication in particular issues and events, Meadows and Avison show that indigenous voices remain infinitely outnumbered by non-indigenous sources in dominant print media in Canada and Australia (2000, p. 348). As Alia adds: "... .indigenous people receive coverage in 'mainstream' media *primarily* in times of crisis ... which encompasses coverage before, during, and after the explosion" (2010, xxiii).

Frustration with dominant media systems, alongside enhanced access to media technologies, has triggered a rise in indigenous media production (Alia, 2010). Taking media into their own hands, indigenous groups have sought to develop narratives that challenge dominant stereotypes. A wave of recent scholarship concentrates on the emergence and growing proliferation of indigenous produced media around the globe (Alia, 1999, 2010; Ginsburg, 2002, 2011; Hafsteinsson & Bredin, 2010; Ledman, 2012; Markelin, Husband, & Moring, 2013; Rodríguez & El Gazi, 2007; Wilson & Stewart, 2008). Ginsburg examines "new aesthetic structures" created by Inuit media-makers to "... recuperate their own collective histories and stories—some of them traumatic—that have been erased in the national narratives of the dominant culture and are in danger of being forgotten in local worlds as well" (2002, p. 40). Rodríguez and El Gazi (2007) describe how indigenous peoples in Columbia use radio to tell their own stories; articulating distinct versions of local realities that stand in contrast to archaic, romanticized and, at times, hostile mediations of their cultures. As Alia summates, "Indigenous-run media do not just fill the airwaves, they carry new worldviews, in sometimes unexpected ways" (2010, p. 4). To exemplify this she describes how Aboriginal Peoples Television Network (APTN) covers weather by placing evenly balanced emphasis on so called remote northern communities and large urban centers in southern Canada.

Indigenous produced media clearly aims to relate more closely to the identities, knowledges and lived experiences of indigenous peoples. In this chapter, we identify some of the steps indigenous activists are taking to reform media coverage of climate change, and strengthen their presence in the news media ecology. But first we must consider the changing nature of media power in the digital age.

Media, Digitalization and Environmental Protest

In *The Whole World is Watching* (1980), Todd Gitlin demonstrates how attempts to attract media attention by the 1960s-era Students for a Democratic Society ended with the movement handing its message over to mainstream media (see also Russell, 2016, p. 144). Building on this classic work, a wide literature has sought to develop a better understanding of the complex interplay between activism, social movements and media power. A relevant strand within this research concentrates on the relationship between news media and environmental protest (Anderson, 1997, 2003; Carvalho, 2005; Hansen, 1993, 2010; Hutchins & Lester, 2006, 2009). With the rise of digital networked communication, focus has shifted from power struggles between environmental activists and traditional news media to questions concerning the changing nature of these power dynamics.

In *Media, Environment and the Network Society*, Alison Anderson raises the question of who holds the power to influence the news media agenda in the network society (2014). One aim of her work is to consider to what extent the digital revolution has altered the balance of media power. She writes that easy access to online connectivity "… through mobile devices radically changes the space of social action and potentially allows it to become increasingly interactive" (2014, p. 6). However, she later adds that "inequalities of visibility" cause "… huge variations across the globe in terms of who has access and the ability to make their voice heard" (2014, p. 6). This paradoxical situation has been at the centre of much contemporary media scholarship and debate.

As environmental activists have gained access to new digital tools, media scholars have turned attention to their evolving media practices (Askanius & Uldham, 2011; Bennett & Segerberg, 2013; Cable, 2016; Dahlberg-Grundberg,

2016; Dahlberg-Grundberg & Örestig, 2016; Melián, 2016; Russell, 2016; Uldham & Askanius, 2013). A relevant strand within this literature examines the interplay between digital media and environmental protest. Dahlberg-Grundberg and Örestig analyze a 2013 anti-mining campaign in Kallak, Sweden that combined local, place-based resistance with social media strategies. This merging of physical and social media activism enabled activists, they argue, to link a physically remote political conflict in northern Sweden to globally contentious mining issues (Dahlberg-Grundberg & Örestig, 2016). In the Latin American context, Melián details how a combination of digital media practices served to enrich and enhance traditional approaches to environmental mobilization (2016, pp. 26–27). She observes, for instance, how mobile smart phones made demonstrations more accessible by providing a system of instant communication that kept channels of communication open between activists based in different locations (Melián, 2016).

Others have concentrated on the role *digital media platforms* are playing for indigenous activists struggling in local contexts to reach wider audiences (Carroll & Lawson, 2016; Coates, 2015; Dahlberg-Grundberg & Lindgren, 2014; Russell, 2005). In the late 1990s, the Zapatista network made unprecedented efforts to create online narratives that re-appropriated dominant cultural myths and stereotypes in order to enhance global awareness of their local struggle against corporate and military power (Russell, 2005). Carroll and Lawson examine how Ojibwe[1] activist bloggers integrated traditional ecological knowledge into their online efforts to protect wolves in the western Great Lakes region of the United States (2016). They write that "… through the strategic use of media technology to provide a platform for tribal resistance, Ojibwe people acknowledged the rights of other-than-human beings as expressed through traditional teachings, or cosmovisions … cosmopolitical work via new media continues to promote indigenous ways of life" (2016, p. 132). In both cases, digital media become tools of resistance used by indigenous activists to develop counter narratives and generate wider support for their causes.

Indigenous Media Vanguard

In late 2012, Idle No More began as local resistance to sweeping legislation passed by the Canadian government. It quickly evolved into a powerful grassroots movement that called "… on all people to join in a peaceful revolution, to honour Indigenous sovereignty, and to protect the land and water" (Idle

No More, 2017). The movement's strategic use of webpages, Facebook groups and Twitter feeds worked in conjunction with the emblematic #idlenomore to consolidate a wide network of indigenous activists, allowing them to articulate a core vision while coordinating efforts and political actions across the country (Coates, 2015). Dahlberg-Grundberg and Lindgren argue that Idle No More serves as an example of emerging "transnational or translocal networked forms of protest" that unite geographically remote social movements with similar agendas (2014, p. 51). However, they are also careful to note that the rising influence of hashtags, and other digital networking tools, raise new questions concerning the social movements they signify (Dahlberg-Grundberg & Lindgren, 2014, p. 71).

Since Idle No More, a number of hashtags have performed a similar function in the North American context. The selection of hashtags below demonstrates how indigenous activists are utilizing digital tools, like Twitter, to amplify local struggles.

- #notyourindigenousstereotype challenges dominant stereotypes of indigenous peoples;
- #nativeappropriation confronts issues of indigenous cultural appropriation in food, fashion, sports, toys and other popular culture (see also Keene, 2010, 2016);
- #changethename and #notyourmascot challenge "mascoting" by professional sports franchises that continue using racialized imagery and language (see also Keene, 2017).

In *Journalism as Activism* (2016), Russell characterizes such digital media practices as the product of an emerging "… news-media vanguard of political activists and innovative traditional reporters who are changing the way reporting the news works to serve the public interest" (2016, p. 8). Indigenous environmental activists are part of this news-media vanguard, as recently exemplified by media coverage of local indigenous resistance to the Dakota Access pipeline development in Standing Rock, North Dakota. As the tension between community members and security forces mounted, activists turned to Facebook Live and Twitter to provide a window into the demonstrations as they unfolded on the ground. Raw video showed security personnel and police using dogs, pepper spray, and a water canon to disperse peaceful Native American protestors. The footage quickly went viral on social media, inducing shock and public outrage while acting as a catalyst for indigenous and non-indigenous peoples from around the world to show their solidarity with members of the Standing Rock Sioux tribe. With the help of digital tools, a local struggle with limited mainstream media exposure was quickly magnified into an international news story. In this instance, the people of Standing Rock helped shape the news media agenda: "… expanding what it means to be involved in the production of news and, in the process, gaining influence over how traditional news stories and genres are constructed and circulated" (Russell, 2016, p. 7).

Breindl writes of the ambivalent relationship between activists and the mainstream media that the primary goal of activism is to change public perception of an issue by reaching the widest possible audience. However, activists are conscious of the professional media's tendency to privilege "… elite voices and render an imprecise picture of reality" (Breindl, 2016, p. 2). In the networked public sphere, Briendl argues that activists must become news producers themselves. To do so, they must:

> … specialize on a given topic, monitoring its evolution through national, supranational and international decision-making processes by using online tools that allow them to connect with like-minded individuals. Activists hold a central position in the networked public sphere, as *brokers* or *bridges* between decision-makers and citizens, similar to the position upheld by media in democracies. (Breindl, 2016, p. 12, emphasis in original)

UN climate summits present a unique moment to examine these dynamics. In the ensuing discussion, we observe a paradoxical situation that transpires at *international* climate summits where *transnational* indigenous activists find themselves marginalized in terms of access, visibility and voice.[2] The question this raises is whether, and to what extent, new media tools are altering the balance of power for indigenous peoples in this particular news media ecology.

Journalism and UN Climate Summits

Climate change in general and the UNFCCC COP summits in particular provide a unique opportunity to study the dynamics of transnational journalism (Eide & Kunelius, 2012a, 2012b; Eide, Kunelius, & Kumpu, 2010; Kunelius, Eide, Tegelberg, & Yagodin, 2016). Climate change is an issue that transcends the boundaries of nation-states. It demands global solutions that must be enacted at the local level. Climate summits grapple with this challenge by bringing together a wide cross-section of actors that includes politicians, bureaucrats, scientists, lobbyists and activists among others. During these events, politics and journalism can in principle be viewed as transnational fields since "… summit actors are (always) situated simultaneously in local and global contexts" (2012b, p. 269). This mixing of scales complicates the role of the professional journalists who cover these "global media events", offering a structurally unique moment to study climate journalism because of the simultaneous coverage produced worldwide (Eide & Kunelius, 2012a, p. 17). Moreover, the media's preoccupation with national and international scales complicates the role of transnational activists who compete to get their messages across in a highly saturated news media ecology.

These transnational conditions and communication challenges are the impetus for MediaClimate, a global network of researchers that has studied climate journalism in over twenty countries since 2007.[3] One of our main findings has been to show that despite the potential for transnational journalistic production to occur, summit coverage is characterized by a considerable degree of nationalistic domestication (Eide & Kunelius 2012a; Eide et al., 2010; Kunelius et al., 2016). In other words, journalists tend to concentrate on the drama unfolding at these summits by filtering it through the lens of national political realities while neglecting the more complex, transnational and global challenges posed by the issue of climate change. Another key finding is that the voices of vulnerable groups, such as indigenous peoples, poor people and women, particularly in developing countries, are marginalized in climate journalism (Eide & Kunelius, 2012a; Kunelius et al., 2016; Roosvall & Tegelberg, 2012). This is explained, in part, by the transnational identities of these groups in media coverage that prioritizes domestic national framings (Eide & Kunelius, 2012a; Eide et al., 2010).

These global findings are mirrored in Canadian and Swedish climate summit journalism. Closer scrutiny reveals that the political game has dominated much of the coverage in our respective nations. When the political game frame is evoked, stories concentrated on the high stakes political negotiations unfolding during each summit, filtering these dynamics through a domestic lens (Roosvall, 2010; Roosvall & Tegelberg, 2012; Tegelberg, 2010).[4] Our previous research on (the limited) coverage of the indigenous activists and leaders who attend UN COPs closely aligns with these broader findings (Roosvall & Tegelberg, 2013, 2015, 2016). In material from Copenhagen and Durban, we found that these actors, and the knowledge they bring to climate summits, are undervalued and underrepresented in mainstream summit journalism. Moreover, in rare instances where mainstream media cover indigenous peoples, they are used to highlight the devastating impacts of climate change (victim frame) or as those who possess knowledge that can help solve the problem (hero frame) (Roosvall & Tegelberg, 2012, 2013). Building on these findings, here we examine how indigenous activists are responding to limited access to mainstream media. Namely, by turning attention and contributing to a growing proliferation of digitally "networked publics" (Russell, 2016).

Legacy Media, Its Centrality and Limitations

There is abundant evidence of the need for stronger representation of indigenous voices in global climate journalism. Our ongoing efforts to map and examine representations of indigenous peoples in UN summit journalism demonstrates their absence from legacy print media across the globe (Roosvall, 2010; Roosvall & Tegelberg, 2012, 2013; Tegelberg, 2010). Table 4.1 provides a summary of the headline data collected by the MediaClimate network during four UN climate summits.[5] It is important to take stock of headlines since they function as a primary means of framing stories about climate change and shaping public perceptions (Antilla, 2005; Entman, 1993; Nisbet, 2009). In spite of a strong indigenous presence at international climate summits, our findings indicate that their knowledge and experiences of climate change are seldom featured in global headlines. This pattern has occurred despite the fact indigenous groups often live in close proximity to the most immediate effects of climate change (Salick & Byg, 2007; Tauli-Corpuz & de Chavez, 2009). By contrast, considerable media coverage has been dedicated to depicting Arctic climate change using powerful visual icons such as polar bears and melting ice caps (Christensen, Nilsson, & Wormbs, 2013).

At COP21, the absence of indigenous voices in global print journalism was palpable. There were only 7 headlines that made reference to indigenous peoples in a sample of 2,223 stories; and only 6 stories in Canada and 6 in Sweden respectively. These findings are mirrored in a recent study on how IPCC assessment reports cover and frame indigenous content (Ford et al., 2016), showing that the voices of indigenous peoples have mostly been ignored in dominant mediated and scientific discourses on climate change.

Table 4.1. Indigenous peoples in longitudinal COP headline data, 2007–2015.

	COP 13	COP 15	COP 17	COP 21	Total
Total headlines	987	3,210	734	2,223	7,154
Indigenous keywords in global headline sample[a]	2	4	3	7	16
# of countries in headline sample	19	19	18	20	N/A

[a]The following keywords were used in each of these searches: "Indigenous", "Aboriginal", "First Nation", "Inuit", "Eskimo", "Native American" "Sami" "Aborigine" and "Māori". There are thousands of distinctive names for indigenous peoples and we recognize that some specific references to these groups may occur in the headlines sampled. However, mining the headline sample for these references was beyond the scope of our present study. What our findings with these broader search terms indicate is that coverage of indigenous peoples and climate change has remained sparse in global media coverage of UN climate summits since 2007. We are also aware that "Eskimo" is a pejorative term, however we selected it as a search term because it continues to be used by news outlets in some parts of the world.

Absence and silence also marked our more targeted efforts to study the representation of indigenous peoples in Canadian and Swedish climate journalism (Roosvall & Tegelberg, 2012, p. 300). Table 4.2 demonstrates the virtual absence of indigenous representation in mainstream print news coverage of UN climate summits in our respective countries. In rare instances where indigenous peoples are featured in this coverage, their primary role is as intermediaries or messengers (Roosvall & Tegelberg, 2013). The dominant representational frame positions them as victim-heroes: *victims*, on one hand, due to their frontlines susceptibility to the devastating effects of climate change; and *heroes*, on the other, when indigenous traditional ecological knowledge is evoked as a possible means of mitigating climate impacts (Roosvall & Tegelberg, 2013). By contrast, the agency of indigenous peoples as *transnational* political actors with a considerable stake in the outcomes of *internationally* sanctioned climate summits has rarely been acknowledged in such reporting.[6]

Table 4.2. Representations of indigenous peoples in Canadian and Swedish story data, 2007–2015.

	COP 13	COP 15	COP 17	COP 21	Total
Canada	3	4	2	6	15
Total stories	92	262	73	145	572
Sweden	0	3	3	6	12
Total stories	54	157	106	200	517

Our quantitative findings were reinforced in the interviews and participant observation we conducted at two UN climate summits. Indigenous activists from the Pacific Islands to the Arctic frequently emphasized the importance of bringing their issues and perspectives to the attention of mainstream national and international media. Yet they unanimously expressed concern with the lack of attention paid to their perspectives in legacy news media coverage of climate change, and summit journalism in particular. Hence, many interviewees discussed their own strategies for grappling with this challenging situation.

At COP17, all seven interviewees spoke of the limited coverage indigenous perspectives have received in the national news media ecologies of their respective countries (Roosvall & Tegelberg, 2016). As Vibeke Larsen put it, "The mainstream media in Norway don't cover indigenous problems anyway … they cover climate change but they don't cover the indigenous aspect in climate change" (Sami, Samiland, Norway, COP17). However, most also stressed that getting the mainstream media to pay more attention to indigenous climate issues was a primary objective. Raymond de Chavez emphasized the importance of *finding a hook* by developing a "good story line that will really catch the attention" (Igorot people, Philippines, COP17). Tiina Kurvits (Many Strong Voices, Canada, COP17) described one of her organization's strategies which entails bringing influential leaders from Arctic indigenous communities and small island states together at press events in order to attract the attention of major international media (BBC, CNN, National Geographic, etc.). She implies that using cultural difference to create a spectacle is necessary, in this case by bringing together leaders representing two distinct local cultures, in order for communication about acute climate change impacts to have broader appeal. Cultural and symbolic imagery is hence what draws media attention, and thereby recognition of identity rather than status.[7]

At COP21, indigenous activists continued to comment on the centrality of legacy news media. India Logan-Riley described how effectively crafting "media spin" can help generate further awareness of indigenous issues:

> Spin is crucial, basically. How the media sees, say, messaging at a protest in a banner, or in a press release, is really important to then how that is communicated to the public and whether you manage to build your movement or isolate your movement. … And also getting media exposure in general, accessing the right avenues and getting it out to the right avenues. (India Logan-Riley, Māori, New Zealand, COP21)

Here some of the burden of gaining media exposure falls on indigenous peoples. Logan later stressed that climate journalism could be enhanced by giving it "a human face", and providing more "context", emphasizing this is what the coverage currently lacks.

Several activists echoed these concerns, elaborating on the mainstream news media's failure to adequately cover climate change and its impacts on indigenous peoples. As Justine Dammay explained:

> Our media in the Philippines is not covering what really is the cause of climate change. They are not covering the cause of why our rivers are being polluted—why our mountains are being … why they're mining them … they are not really focusing on that issue. It's like they're focusing on the cost of that individual person, but they do not focus on the big cost, or big problem—on the bigger picture. (Justeene Dammay, Tuwali people, Philippines, COP21)

Two interviewees from New Zealand and the Pacific Islands expressed similar concerns.

> I guess most of my experience comes from New Zealand coverage and it tends to be very flash in the pan, short, sharp, and then what's the next big scandal. Consistent, comprehensive, and thorough investigative reporting isn't in mainstream New Zealand media on the TV and in newspapers … there isn't much done in terms of holding the politicians to account and giving them a bit of a blasting for not taking any action. (India Logan-Riley, Māori, New Zealand, COP21)

> [C]limate change is actually pretty … you know, it's new to our ears, so I think the media could do a bit more … like having more information on what is climate change and how is it caused … how does it impact the lives of [indigenous] peoples, and how it can be stopped. (Mikaele Maiava, Tokelauan, Tokelau, COP21)

Without local, investigative journalism, Logan-Riley stressed that there was no way to hold officials accountable for their inaction on climate change. Moreover, for Maiava, the mainstream media's failure to raise awareness of how climate change is impacting indigenous communities—at the local level and beyond—must be addressed. In each instance, tension emerges between the centrality and limitations of media coverage of climate change and indigenous peoples at the local, national and international scales.

Three indigenous interviewees from Russia expressed similar sentiments about the news media ecology. They unanimously felt that Russian news media coverage of climate change has been limited at best (Dmitry Berezhkov, Itelmen people, Olga Atsynga Letykai Csonka, Chukchi people, Victoria Sharakhmatova, Itelmen people, Russia, COP21). This is supported by recent research on the low volume of coverage of climate related news in Russian print media (Poberezhskaya, 2015; Yagodin, 2010). An interviewee from Siberia linked this broader issue to coverage of Russian indigenous peoples: "… [the] media manipulates the subject of climate change and I would like to see that it is not manipulated, especially when it comes to the questions of indigenous peoples" (Victoria Sharakhmatova, Itelmen people, Russia, COP21) Jannie Staffansson described similar issues with the accuracy of reporting on Sami issues in Scandinavian media. Consequently, she pays more attention to international media: "Sami people, we have had a better time with international media so I do tend to follow the bigger news from international media. … Even though sometimes there are people reporting on Sami or indigenous peoples in media, it is not correct media" (Sami, Samiland, Sweden, COP21). One means of responding to manipulation and misrepresentation, it seems, is to turn to another part of the media ecology where journalists are more receptive to indigenous issues.

When asked how to remedy these issues, many interviewees called for mainstream media to pay closer attention to indigenous peoples, and their knowledge of climate change. Anthony Debbarma made a broad call for the inclusion of more indigenous voices in global media coverage. He felt that more media exposure would:

> … create awareness … [so] the government will understand what we are doing, because if the media doesn't report then governments will not see … So if media started covering what the indigenous people or what other people are actually saying about climate change, then they—the government, the negotiators—will see what is happening *outside* of that room. (Anthony Debbarma, Borok people, India, COP21)

Lene Kielsen Holm expanded on how further inclusion of what indigenous peoples have been observing locally can enhance global public awareness of the acute effects of climate change.

> … indigenous peoples from around the world have been observing changes that they had not seen before and rely on these findings. And maybe already fifteen, twenty years ago, they [governments] could have invested in trying to reduce their emissions and the result would have been a little better than the one that we can expect in five to ten years. (Lene Kielsen Holm, Inuit, Greenland, COP21)

These sentiments were echoed by two Inuit youth representatives from the Canadian Arctic. Jordan Konek said:

> We're here to talk about our elders' perspective on climate change and hope that the government or the higher powers will listen to our elders' perspective … they have been out on the land for so long and they see the changes and they know the changes and that is what the government should look at and study … [the elders'] perspective on climate change because they know more than the scientists. (Jordan Konek, Inuit, Canada, COP17)

It is not only indigenous perspectives on climate change but also a general knowledge of indigenous peoples that is sorely lacking in mainstream media. As Vibeke Larsen put it "… they don't know the indigenous people. It seems like media don't have knowledge about all the indigenous people in the world, the overall media. It seems like lack of knowledge" (Sami, Samiland, Norway, COP17). After expressing a similar view, Chukchi artist Olga Atsynga Letykai Csonka invited media to "come and see them [indigenous peoples], how they live" (Chukchi people, Russia, COP21). Like Larsen and Csonka, Allison Akootchook Warden, stressed that coverage would improve if journalists started "… travelling to the [indigenous] communities that are most impacted by climate change, I think, to get the story firsthand" (Iñupiaq people, USA, COP21).

In our participant observation at the COPs, we witnessed legacy media's marginalization of indigenous peoples play out first hand. As we have argued elsewhere, a geography of injustice permeated the summit space, with indigenous spokespersons, desiring to participate in the politics and mediation of these events, relegated to venues distant from the official proceedings and from the core media hub.[8] At COP17 in Durban we sat in press conferences, held by indigenous organizations, that were sparsely attended and where few, if any, questions were asked by the journalists present. At COP21, we attended events that attracted more media attention. However, these events all took place at venues far away from the highly regulated spaces inside the summit where the major media players were mainly to be found (see Chapter 3).

Writing more broadly on activists as news producers, Breindl notes that, "… reaching mainstream media remains crucial for activists, if they wish to reach concrete changes. The Internet may have decreased their dependence on the media but not cancelled it" (2016, p. 12). Interviewees reflected this in their demands for mainstream media to pay more attention to their concerns; even with the enhanced access and amplified voice social and mobile media may afford them. The statements examined here illuminate the paradoxical situation indigenous activists find themselves in at UN COPs. On one hand, it is evident that they continue to view legacy media as a crucial channel for disseminating their voices and experiences of climate change. On the other hand, the examples attest to a collective frustration with the limited attention indigenous experiences of climate change receive in mainstream media.

To find out how indigenous activists grapple with this paradox, we now turn to some of the strategies they employ to counter underrepresentation in the legacy news media ecology. We identify two main responses: firstly, *turning to social media* as an alternative source of pertinent news and information; and secondly *producing their own media*, using digital tools to facilitate the self-representation of indigenous climate issues.

Social Media as News Source and Connective Tool

In our media monitoring of summit journalism, we have witnessed a gradual turn toward social media in indigenous efforts to engage with the public and the UNFCCC process since 2007. This is explained, in part, by a general rise in the popularity of social media platforms over the same time period. However, we also view this as one of the ways indigenous activists are responding to underrepresentation in legacy media. The rising influence of social media became apparent in the interviews we conducted at COP17 in Durban (2011). Activists representing several organizations (Tebtebba, Inuit Youth Delegation, Many Strong Voices) identified with social media as a new means of accessing and disseminating critical information, frequently neglected by mainstream media. Moreover, these actors stressed that social media was helping to highlight the *political* role of indigenous groups at the summit.[9] By COP21 (2015), interviewees consistently identified Facebook and Twitter as primary sources of general climate news, and for news concerning particular events unfolding during the COPs. Social media news was praised for its raw, unedited quality in contrast to mainstream news coverage dominated by spin and manipulation.

This positive appraisal of social media news took shape in two main ways. The first was by interviewees who identified Facebook or Twitter as primary news sources. Justeene Dammay compared reading mainstream news "… maybe five times a week—not really daily" with checking her Facebook feed for news "five times a day" (Tuwali people, Philippines, COP21). Another interviewee described her news consumption on Twitter: "I have 5,700 Twitter followers. So I'm a little addicted to Twitter … I get most of my news from Twitter, and then also news from friends. I don't really follow the news outside of that, but I have a pretty good system of communicating on Twitter" (Allison Akootchook Warden, Iñupiaq people, USA, COP21). For Warden, Twitter serves the dual purpose of functioning as a news source and a tool for communication.

Another indigenous activist with the Pacific Indigenous Network distinguished between news coverage in legacy and social media as follows: "…

with Facebook … you get raw materials. But with the [mainstream] others, sometimes you are not even covered because you are saying something that is not politically right, you know […] social media covers … a lot more because … you know *everybody … can be a journalist.*" (Mikaele Maiava, Tokelauan, Tokelau, COP21, emphasis added). Facebook news is viewed as ideologically more honest, well-rounded and trustworthy than what is considered ideologically biased coverage found in legacy media.[10] Others expressed related sentiments to explain why they never watch television.

> I don't own a TV. I think TV is bullshit to be honest, I don't watch TV, you know. So unless someone shares it or sends it to me then I don't watch it … people that I follow on YouTube, people I follow on Twitter, and friends of mine on Facebook really, that's where I get all of my news to be honest. (Teanau Tuiono, Māori, New Zealand and Cook Islands, COP21)

Sami Council President Áile Jávo observed a growing shift away from traditional media:

> Social media has become … very important over the last few years … starting to overthrow, so to speak, the traditional media … Links that are being shared on Facebook and Twitter and so on … that's what people read. They might not go to the main page in the diverse media. In that sense, social media are very important … moreover you can share things there that traditional media do not cover. But it varies of course … the quality … and how sustainable the information there is. (Áile Jávo, Sami, Samiland, Norway, COP21)

Sami journalist Pirita Näkkäläjärvi, Head of Yle Sápmi (the indigenous division of the Finnish national public service broadcasting), described how her organization uses social media as "… a major distribution channel for our content in Finland, Sweden, Norway and Russia. Social media is used every weekday." She then expanded by providing several examples of how her organization uses social media to provide live coverage:

> We do livetweeting from major events on Twitter. We post photos from cultural events in Facebook and Instagram. Snapchat we use more haphazardly to create buzz for younger audiences. Some of our journalists post their stories in specific Facebook groups active around for example the Inari and Skolt Sami languages. Our journalists are daily contacted by people on Facebook chat. People give feedback and send news tips. (Pirita Näkkäläjärvi, Sami, Samiland, Finland, Head of Yle Sápmi, e-mail interview 2017)

Among the indigenous activists and leaders we interviewed, social media is evidently replacing mainstream media as a primary source of climate journalism. This entails placing more trust in media produced or shared by friends and colleagues than in mainstream climate news produced by professional journalists. Maiava's comment, that anyone with access to social media can now be a journalist, resonates with broader debates concerning the changing nature of journalism (Fenton, 2009). As global publics increasingly place more trust in citizen journalism, professional journalists must find ways to shape up. Several interviewees, as previously noted, felt that one way to do so would be for journalists and mainstream media outlets to invest more resources in covering climate change in frontline indigenous communities.

Social media were also frequently identified as useful platforms for following the COPs, especially for keeping up with the latest news on indigenous events and actions. Jannie Staffansson identified Twitter as a key source of COP news: "I do also follow Twitter, especially during these sessions, because a lot of my friends are on Twitter so you can get information about what's happening during COP" (Sami, Samiland, Sweden, COP21). India Logan-Riley underlined that youth and officials are increasingly turning to social media for news during the summits.

> … it is how COP tends to function I've noticed, especially in the younger constituencies. I think also negotiators tend to follow it a bit more, Facebook, social media, stuff like that because it can give you a broad spread of what the general feeling is about a certain thing. … yeah, social media is a big deal, and because *a lot of people have disengaged with mainstream media and have gone to social media* it can be a better way of accessing, say, your peers and sharing stuff with them. (India Logan-Riley, Māori, New Zealand, COP21, emphasis added)

The examples suggest that social media is quickly supplanting legacy media as a dominant source of news and information for activists at the COPs. Hence, the next section takes a closer look at how indigenous issues were taken up on Twitter during COP21.

Social Media as Connective Tool

To capture a snapshot of indigenous activity on Twitter, we used a digital research tool to harvest indigenous Twitter data during COP21.[11] The resulting sample consists of 1,265 tweets that used #indigenousCOP21, involving 684 unique user accounts.[12] Our search time frame was limited to tweets published during the final days of the summit between December 8 and December 15, 2015. The data collected includes details on the content of the tweets (words, @mentions, URL links, hashtags) as well as account details and statistics on the Twitter users included in the sample. We realize this is a very small number of tweets and active users relative to the 4.6 million tweets that featured #COP21 during the summit (Sylvestre, 2016). However, the purpose of examining this small cluster of tweets is to consider the extent to which Twitter, and other social media platforms, function as a means of building networked publics where indigenous actors and other concerned parties could exchange information and opinions on relevant events as they transpired at COP21. This digital network, however small, represents an alternative to a mainstream media ecology where indigenous peoples have largely been marginalized and ignored. By taking a closer look at active tagging and sharing practices, we can identify some features of the indigenous networked public that took shape during the summit; a network wherein indigenous peoples clearly sought to communicate within, across and beyond their localized spheres of influence.

Table 4.3 lists the top ten hashtags in the #indigenousCOP21 data sample. What stands out is that more than half of these tweets (56%) include #COP21. Hence, a large portion of the tweets in this sample were constructed with the intent of communicating with everyone at COP21 (and not just with indigenous peoples). While it is impossible to determine the success of these efforts, the result suggests that tweets in this network frequently targeted a wider audience than those already conversant with indigenous issues. For example, #indigenous and #climatechange accounted for 21 percent of the total number of hashtags mentioned in the sample. This demonstrates how hashtags can be used to link particular perspectives and struggles to broader

Table 4.3. Top ranking hashtags in #indigenousCOP21 Twitter data.

Top 10 hashtags	% of total
#indigenouscop21	100
#cop21	56
#indigenous	14.15
#climatechange	7.4
#arctic	7
#humanrights	4.3
#nncs	2.8
#norden	2.8
#glpol	2.8
#dontdamtheclimate	2.8

debates pertaining to climate change and indigenous issues. Additionally, a cluster of regionally specific hashtags accounted for 15 percent of the hashtags sampled (#arctic, #glpol—Greenland politics; #norden—Scandinavian reference to the Nordic countries; and #nncs—New Nordic Climate Solutions; a Nordic political organization). This tendency suggests, yet again, that tweets were frequently crafted with the intent of reaching beyond a localized echo chamber (Williams, McMurray, Kurtz, & Lambert, 2015) to communicate with wider publics. In this instance, indigenous issues are linked to broader discussions concerning climate politics in the Arctic, Greenland, and the Nordic regions. Finally, although climate justice was not a prominent hashtag in the sample, 4.3 percent of tweets used #humanrights, linking indigenous peoples, climate change and the Paris COP to the broader issue of human rights, thereby connecting to discourses of climate justice.

Table 4.4 shows a strong presence of transnational indigenous networks in the top ranking user mentions in the Twitter sample. The networks included among the top @mentions are: International Indigenous Peoples' Forum on Climate Change (@iipfcc); Amazon Watch (@amazonwatch); Indigenous Peoples' International Centre for Policy Research and Education (@tebtebba); Indigenous Environmental Network (@ienearth); and Asian Indigenous Peoples Pact (@aippnet). Each of these networks has maintained a strong presence at the COPs, in spite of the little coverage they receive in mainstream media. Twitter, and other social media platforms, help to fill this communication void by enabling these actors to communicate with one another, and to contribute to the broader media ecology that surrounds these summits.

Table 4.4. Top ranking @mentions in #indigenousCOP21 Twitter data.

Top 10 @mentions	#
@350europe	88
@iipfcc	53
@amazonwatch	32
@tebtebba	28
@cfrontlines	28
@arcticjourno	27
@ienearth	26
@gridarendal	24
@ineqikielsen	23
@aippnet	23

It is noteworthy that 350.org Europe (@350europe) received the highest number of mentions in the sample. In a social network analysis of 4.6 million #COP21 tweets, Sylvestre (2016) found that 350.org was a central node in the network of environmental activists and NGOs that formed in the days leading up to the Paris Agreement. The high frequency of 350.org mentions in the #indigenousCOP21 tweets suggests that this was one way indigenous activists sought to link their tweets—and corresponding demands—to broader conversations concerning the Paris Agreement. The Indigenous Environmental Network (@ienearth) is another central node in Sylvestre's (2016) social network map of the wider community of activist and NGO tweets. @ienearth's centrality in this expansive map of #COP21 tweets infers that indigenous activists had some success bringing their issues to the attention of a wider audience on Twitter.

Table 4.5 shows the top ranking media sources that were linked to the dialogue surrounding indigenous COP21 on Twitter. It shows the diverse range of voices, perspectives, and regions of the world represented in this online discourse. This differs from mainstream media where the voices are predominately from the professional, political realms and, as we demonstrated earlier, rarely include indigenous peoples. By contrast, we have no way to track Twitter users who discussed indigenous issues without using #indigenousCOP21. Hence, it becomes easy to overlook critical information or key actors due to hashtags that can fragment the conversation by dividing it up into separate spheres.

Table 4.5. Top ranking sources in #indigenousCOP21 Twitter data.

Top 10 sources	#
UNFCC Paris Agreement	17
Inuit Circumpolar (Arctic)	10
The Guardian (UK)	9
UNFCCC Draft Paris Outcome	8
Indigenous Climate Portal	7
National Observer (Canada)	7
Asia Indigenous Peoples Pact	7
YouTube Chirapaq (Peru)	6
IIPFCC Press Conference	5
InterAksyon (Philippines)	5

The volume of tweets, and the diversity of actors, reveals that indigenous perspectives were indeed present on Twitter at COP21. Analysis of the

#indigenousCOP21 tweets reveals that this hashtag was frequently combined with other, climate-oriented hashtags—most notably #COP21—that at least had the potential to reach larger communication networks engaged in climate politics. Thereby we can assume that indigenous peoples used Twitter to speak out and encourage other people to listen. However, the Twitter data leaves many questions unanswered. For one, it is difficult to determine the scope of the #indigenousCOP21 communication network. Was this an open forum (Williams et al., 2015) where indigenous voices contributed to wider debates unfolding at COP21 by amplifying locally oriented perspectives? Or was this a "community cluster" (Smith, Rainie, Himelboim, & Shneiderman, 2014) that primarily consisted of actors with a common interest in indigenous issues?

Becoming the Media: Multifaceted Strategies of Indigenous Self-representation

Another strategy for responding to underrepresentation has been for indigenous activists to produce their own media. Breindl writes that "… to bring about political change, activists need to monitor, produce and distribute political information that is often neglected by traditional media outlets. They can use various protest channels to do so" (2016, p. 6). To fill the legacy media void at climate summits, transnational indigenous groups send representatives who serve as *knowledge brokers* between local constituents and the civil society actors on hand at these events (media, politicians, etc.). By facilitating the establishment of "networked publics" (Russell, 2016), digital media offer these actors new and diverse channels of self-expression where they can produce and disseminate narratives that transcend the instrumental nature (i.e. a preoccupation with political games) and dominant framing tendencies of mainstream summit journalism. In sum, indigenous activists are increasingly carving out their own spaces of representation in the news media ecology to overcome marginalization.

The need for self-representation of neglected voices and perspectives was identified as a motivating factor for many of the indigenous activists we interviewed. At both summits, they discussed their role as communicators, outlining two main purposes for their media practices. Firstly, to communicate *local* perspectives on the impacts of climate change with *the rest of the world*. We refer to this as the local-global connection. Secondly, to communicate and translate information gathered at *global* climate summits with *local constituents*. We refer to this as the global-local connection. Activists outlined communication strategies that spanned from efforts to engage traditional media to producing their own media using a range of different digital tools and platforms.

At COP17, five interviewees said that enhancing global awareness of local climate impacts was a primary objective. Raymond de Chavez (Igorot people, Philippines, COP17) described a daily routine *inside* the Durban summit that included meeting with the Indigenous Caucus to develop a "strategic communication plan", organizing press conferences, doing interviews with journalists, and minding an information booth in the poster room. He also mentioned the growing importance of Twitter, Facebook, YouTube and Skype for internal communication within his network. He credited these social networking platforms with helping to alleviate the high costs associated with travel in order for network members to meet in person.

Frustrated with the corporate controlled mass media, Francois Paulette (Dene, Canada, COP17) prioritized working *outside* the summit with activist groups, such as the Indigenous Environmental Network, to heighten public awareness of the devastating impacts of the Tar Sands. Tito Puanchir (President of CONFENIAE) described the multifaceted strategy his organization uses to translate and transfer global climate knowledge for the vast network of Amazonian indigenous communities they represent.

> We plan assemblies, congresses, as a way to spread the information, sometimes we invite people from other places who know the situation and they help us with the information, amplifying it and spreading it with different media, writing, images, ways that can help people at the local level, at the level of the communities to understand what is happening. (Tito Puanchir, Shuar people, Ecuador, COP17)

He also mentioned the importance of community radio for disseminating information across the expansive territory represented by his organization. Hence, the choice of media can vary according to contextually specific conditions of media access at the community level.

Seven interviewees said their main purpose for attending COP21 was to do communication work for one or more indigenous organization. Dmitry Berezhkov said his main role at COP21 was to represent Russian indigenous peoples (Itelmen people, Russia, COP21). During the regional day for Russia and Eastern Europe (December 4), he was listed as a media contact for all of the scheduled events. He kept busy conducting interviews with reporters throughout the day. Jon Petter Gintal detailed his communication work as follows: "I am the contact person for the media for the Sami parliament [in

Norway], so we had a press release when the COP started, and I had three or four different media that I have been in touch with, including the Sami radio" (Lule Sami, Samiland, Norway, COP21). Later, he mentioned Twitter: "I know the President of the Sami Parliament—she's very active on Twitter. So I know that our communication unit helps her out with that."

After identifying himself as a spokesperson for Ngāti Hine (a Māori organization), Teanau Tuiono described a communication strategy that entailed using all forms of media to "… be as visible as possible and to … get as much media attention for the things that we are pushing *inside and outside* … working with indigenous media … with any media that is actually interested in what we are doing" (Māori, Cook Islands and New Zealand, COP21). Mikaele Maiava talked about the importance of indigenous produced messages: "… in our organization we have media people and they do pretty much everything for us. All I do is just type it on my Facebook and share it, or upload it on YouTube and share it. But everything that has to do with media is done by our media people or communications people" (Tokelauan, Tokelau, COP21). He later stressed that: "without the media and, you know, the social media … it would be really hard for our voice to be heard … that is like our vehicle, you know, to ensure that what we are doing, our struggle … is being heard all over the world."

Allison Akootchook Warden talked about doing communication work for three different organizations (It Takes Roots to Weather the Storm, Indigenous Environmental Network and Sierra Club).

> I use a lot of social media because my people are unique, and I think we have a unique perspective on the world and a lot to offer, a lot of wisdom to offer on what the planet is going through right now. So even though it's against my own cultural norms to put myself out there in that way … you know, to have such a big media presence, or to, you know, really put our voices out there, but me, I feel that it's important to do that. (Allison Akootchook Warden, Iñupiaq people, USA, COP21)

For each of these actors, self-representation acts as a vital tool for conveying indigenous knowledge and experiences of climate change to wider audiences. Efforts are made to target and produce media for all different parts of the news media ecology.

Other interviewees spoke about the importance of attending COP21 in order to relay critical information to community members back home. Marie-Josee Artist described a daily routine that entailed communicating on behalf of indigenous peoples in Suriname at climate conferences, while simultaneously reporting on developments in the UNFCCC process to the indigenous communities she represented. She described what can make this a challenging task:

> Our problem is that we work with the communities … meaning that the rest of the population in Suriname, do not hear anything about us. So, we need our website to go out to have a broader reach, so that we have a platform in Suriname and also for non-indigenous peoples because we need that support so that's why we are now working on our website. (Marie-Josee Artist, Kalina people, Suriname, COP21)

Artist wants her message to reach everyone in Suriname, not just like-minded members of indigenous communities. This effort entails producing content for different parts of the media ecology (legacy, digital, indigenous), and for different types of audiences—local, national, global.

At a press briefing on Forest Day (December 1), with a panel of spokespersons from four different continents, we listened as Diel Mochire gave a short presentation and took questions from journalists on behalf of the Network of Indigenous and Local Communities for the Sustainable Management of Forest Ecosystems in Central Africa (REPALEAC). In an interview later, he stressed that speaking to the media is necessary to inform decision-makers and the public about what is happening in forest ecosystems (Pygmy, Democratic Republic of Congo, COP21). Mochire identified an organizational website and a Facebook page as two primary means of communicating about the summit with local community members. However, he added that issues of access, a lack of resources and technical fluency, make it difficult for community members to use social media, or any media, for the purpose of self-representation: "There is not much follow up on this because they do not have, yeah, it is not very active because they do not have the resources, or technical skills." While this did not stop Mochire from "… using media in all his activities," he emphasized that word of mouth remains the primary means of communicating about climate change locally.

Lene Kielsen Holm described language barriers that can make it difficult for local constituents to follow media coverage of the COPs.

> In Greenland our first language is Greenlandic and the media, most of it is in other languages than Greenlandic. So it's hard for people around Greenland to understand what is going on. So persons like me and others who have some insight on what is going on, *we have to be here to report back to our own people to make them aware of what is happening*. (Lene Kielsen Holm, Inuit, Greenland, COP21, emphasis added)

Social media, she added, is making it easier to disseminate information in local languages: "… this morning I wrote in Greenlandic and English saying

… what the big issues are today for indigenous peoples." Having access to the summits, and to digital tools for indigenous self-representation are both essential ingredients for overcoming language barriers in order to circulate pertinent information at the local level.

Finally, several interviewees described the particular ways in which they used social media channels to communicate with community members at home. Jannie Staffansson talked about using social media as a tool to connect with her community during the climate conference.

> I try to use it [social media] … to inform, mostly for my people back home, so they will have information on what I'm doing on their behalf, and in the name of the Sami people because that is important. … I need to feel that I have their support in what I am doing, what I think and many of us think is the right thing, but it's important to inform them of what we are doing. (Jannie Staffansson, Sami, Samiland, Sweden, COP21)

She added that social media can be used to help engage indigenous youth: "That's what the youth are reading and that information is important for them to reach them. So reaching youth and making them engaged in this is very important."

India Logan-Riley expanded on this by detailing how she used social media to connect with Māori youth who could not attend the summit. To do so, she participated in daily "Climate Hangouts": "… 15 minutes to half an hour update for those who are in New Zealand … we get guest speakers on to talk about what's happened that day at COP, what they are following, what are their experiences, just so that there's a day to day update" (India Logan-Riley, Māori, New Zealand, COP21). These "Climate Hangouts" on social media are designed with the intention of keeping Māori youth informed on the latest developments in Paris. She later mentioned a peer who used his video-blog for the same purpose. Justeene Dammay described an online signature campaign designed by her organization to engage indigenous youth in the Philippines. She stressed that using online platforms made it easier to find support and "…share our issues with the youth—or the other people who use alternative media" (Justeene Dammay, Tuwali people, Philippines, COP21).

Jon Petter Gintal talked about using social media to overcome geographical barriers that make it difficult for indigenous groups to communicate and organize politically.

> I think people use social media more and more, and it's becoming an important part of politics. So I see, especially in the Sami people—we are very few, and we live in a huge area, so then social media plays a good role to keep the communication between

> each other in a good way. So we use social media a lot. And I see also our political level use it … when they have new political points. They publish that on social media. (Jon Petter Gintal, Lule Sami, Samiland, Norway, COP21)

The communication challenges and strategies outlined here entail finding ways to bridge the flow of information between the global and the local and vice versa. Multiple geographical scales are regularly evoked—local, national, intranational, global—to describe the difficult task of trying to engage with local constituents and wider audiences simultaneously.[13] At times it is necessary to liaison with legacy media to support these efforts, while it is also deemed essential to take media production into your own hands. Evidently, these actors consider themselves transnational *knowledge brokers*, tasked with making connections between the global and the local, and between different components of the news media ecology.

Conclusion

This chapter underlines a tension between demands for an amplified voice within the news media ecology and a mainstream media system that frequently silences indigenous peoples. At UN climate summits, indigenous activists perform a balancing act by vying for media attention while trying not to compromise the messages they carry on behalf of local constituents. These actors take *responsibility* for representing the acute climate challenges they face in different parts of the media ecology. Doing so entails carving out new channels of access for voices that come from the *outside*, in order to transform a news media ecology which currently renders these transnational actors invisible.

The indigenous activists we interviewed related to the news media ecology in three main ways. Firstly, by identifying the central role *legacy media* continues to play in bringing local climate issues to the attention of global audiences and influential decision makers. However, these sentiments were tempered by a growing frustration with the limited attention and access mainstream legacy media have given to indigenous climate issues around the globe. The second way the news media ecology was evoked was in a show of much greater enthusiasm for *social media*. Indigenous activists are increasingly turning to these platforms as alternative news sources and new channels for self-expression. However, this enthusiasm for "citizen journalism" and digital democracy should be restrained in light of recent findings concerning the drawbacks of social media for vulnerable communities, particularly with regards to surveillance (Fenton & Barassi, 2011; Stalder, 2012; Treré, 2016). Finally, interviewees reflected on their own roles as producers within the news media ecology. They articulated multifaceted strategies of self-representation, positing themselves as knowledge brokers with the *responsibility to establish connections* between different media forms (legacy, digital, social, indigenous), between different media actors (politicians, journalists, citizens), and between different media spaces (inside and outside the summit halls).

But what about the responsibility of other actors, such as politicians and media professionals? With stronger positions of power in the news media ecology, we have argued these actors hold greater responsibility for justice (see also Silverstone, 2007, p. 153; Young, 2013, p. 147). One crucial aspect of their responsibility is to explain the issues of climate and justice adequately and knowledgably while connecting them to each other. In accordance with Article 19 of the Universal Declaration of Human Rights (UDHR), another crucial aspect is to not merely provide adequate information for people to "receive", but to establish a public space that grants people the right to "… *impart* information and ideas through any media and regardless of frontiers" (United Nations, 1948, emphasis added). Thus politicians and the media are responsible for securing spaces where transnational groups, such as indigenous peoples, can be widely heard without placing too much responsibility on indigenous actors to carve out these spaces for themselves.

A third crucial aspect of the responsibility of politicians and the media for climate justice concerns the last few words in the quote above: "regardless of frontiers". The UDHR makes it clear that such responsibility must be transnational in scope, coinciding with Young's (2013) social connection model of responsibility for justice. Hence, the media should hold politicians responsible for acting on climate change in a transnational way, taking measures that are fair to *all* peoples. Confronting politicians remains strongly tied to professional journalism, underlining the necessity for such a practice, even at a time when citizen journalism has proven to be capable of incorporating more alternative voices. The media should at the same time be held responsible for enabling indigenous actors to convey information about climate change that is *political* in nature.

Attempting to locate indigenous peoples within the news media ecology has illuminated *(dis)connections* between mainstream, digital and indigenous produced media. While they have different roles to play, mainstream media could use a growing proliferation of indigenous produced media to start telling stories that make room for new voices and perspectives on how to confront the challenges posed by climate change. Hence, we now turn our attention to the stories transnational indigenous activists have been telling at the COPs as part of their *political* struggle for climate justice.

Notes

1. The Anishinaabeg (also known as Ojibwe) are indigenous peoples who inhabit a large territory that spans across the border between Canada and the United States.
2. For further discussion of these three types of media marginalization, see Chapter 5.
3. For a full list of the Media Climate network's publications, see https://mediaclimate.net/publications.
4. For further examination of the political game frame and national domestication, see Chapter 5.
5. The headlines were translated into English by coders from each of the participating nations.
6. For further examination, see Chapter 5.
7. For further development of this line of argument, see Roosvall and Tegelberg (2012, 2013) and Chapter 6.
8. For further examination, see Chapter 3 and Roosvall and Tegelberg (2015).
9. For further examination of this political role, see Chapter 5.
10. For further discussion of the differences between un-edited and self-edited media witnessing, see Chapter 5. We expand on some of the pitfalls of placing so much trust in social media later in this chapter.
11. NodeXL is a free, open source software tool designed to assist with gathering, visualizing and analyzing data from different social media platforms. Once the social media data to be harvested is identified, NodeXL imports the material into an Excel spreadsheet where it can be divided into a series of worksheets designed to analyze the material in different ways. The results displayed in Tables 4.3 to 4.6 were created using the NodeXL tool.
12. Tweets that mention several users are represented by separate rows within the data set. Hence, the total number of tweets is representative of the number of interactions that occurred within the network of Twitter users who engaged with the hashtag #indigenousCOP21.
13. For further examination of geographical scales, see Chapter 3 and Roosvall and Tegelberg (2015).

References

Alia, V. (1999). *Un/covering the north: News, media and aboriginal people*. British Columbia: University of British Columbia Press.

Alia, V. (2010). *The new media nation: Indigenous peoples and global communication*. London: Berghan Books.

Anderson, A. (1997). *Media, culture and the environment*. London: Routledge.

Anderson, A. (2003). Environmental activism and news media. In S. Cottle (Ed.), *News, public relations and power*. London: Sage.

Anderson, A. (2014). *Media, environment and the network society*. London: Palgrave.

Anderson, M., & Robertson, C. (2011). *Seeing red: A history of natives in Canadian newspapers*. Manitoba: University of Manitoba Press.

Antilla, L. (2005). Climate of scepticism: US newspaper coverage of the science of climate change. *Global Environmental Change, 15*, 338–352.

Askanius, T., & Uldham, J. (2011). Online social media for radical politics: Climate change activism on YouTube. *International Journal of Electronic Governance*, 4(1), 69–84.

Bennett, W. L., & Segerberg, A. (2013). *The logic of connective action: Digital media and the personalization of contentious politics*. New York, NY: Cambridge University Press.

Breindl, Y. (2016). Activists as news producers. In T. Witschge, C. W. Anderson, D. Domingo, & A. Hermida (Eds.), *The SAGE handbook of digital journalism* (pp. 250–265). London: Sage Publications.

Cable, J. (2016). *Protest campaigns, media and political opportunities*. London: Rowman & Littlefield International.

Carroll, C., & Lawson, A. (2016). New media, activism, and indigenous environmental governance: politics and the Minnesota-Wisconsin wolf hunt. In S. Monani & J. Adamson (Eds.), *Ecocriticism and indigenous studies: conversations from earth to cosmos* (pp. 119–135). London: Routledge.

Carvalho, A. (2005). Representing the politics of the greenhouse effect: Discursive strategies in the British media. *Critical Discourse Studies*, 2(1), 1–29.

Christensen, M., Nilsson, A. E., & Wormbs, N. (Eds.). (2013). *Media and the politics of Arctic climate change: When the ice breaks*. Basingstoke: Palgrave.

Coates, K. (2015). *#IdleNoMore and the remaking of Canada*. Regina: University of Regina Press.

Dahlberg-Grundberg, M., & Lindgren, S. (2014). Translocal frame extensions in a networked protest: Situating the #IdleNoMore hashtag. *IC – Revista Científica de Información y Comunicación, 11*, 49–77.

Dahlberg-Grundberg, M. (2016). *Digital media and the transnationalization of protests* (Doctoral dissertation). Umeå University, Sweden.

Dahlberg-Grundberg, M., & Örestig, J. (2016). Extending the local: Activist types and forms of social media use in the case of an anti-mining struggle. *Social Movement Studies*, 1–14. doi: 10.1080/14742837.2016.1268955

Eide, E., & Kunelius, R. (Eds.). (2012a). *Media meets climate: The global challenge for journalism*. Gothenburg: Nordicom.

Eide, E., & Kunelius, R. (2012b). Moment of hope, mode of realism: On the dynamics of a transnational journalistic field during UN climate change summits. *International Journal of Communication*, 6, 266–285.

Eide, E., Kunelius, R., & Kumpu, V. (Eds.). (2010). *Global climate, local journalisms: A transnational study of how media make sense of climate summits*. Bochum: Projekt Verlag.

Entman, R. M. (1993). Framing: Toward clarification of a fractured paradigm. *Journal of Communication*, 43(4), 51–58.

Fenton, N. (2009). *New media, old news: Journalism and democracy in the digital age*. London: Sage.

Fenton, N., & Barassi, V. (2011). Alternative media and social networking sites: The politics of individuation and political participation. *The Communication Review, 14*(3), 179–196.

Ford, J., Cameron, L., Rubis, J., Maillet, M., Nakashima, D., Willox, A., & Pearce, T. (2016). Including indigenous knowledge and experience in the IPCC assessment reports. *Nature Climate Change*, 6, 349–353.

Fraser, N. (2008). *Scales of justice: Reimagining political space in a globalizing world*. New York, NY: Columbia University Press.

Ginsburg, F. (2002). Screen memories: Resignifying the traditional in indigenous media. In F. Ginsburg, L. Abu-Lughod, & B. Larkin (Eds.), *Media worlds: Anthropology on new terrain* (pp. 39–57). Berkeley, CA: University of California Press.

Ginsburg, F. (2011). Native intelligence: A short history of debates on indigenous media. In J. Ruby, & M. Banks (Eds.), *Made to be seen: A history of visual anthropology* (pp. 234–255). Chicago, IL: University of Chicago Press.

Gitlin, T. (1980). *The whole world is watching: Mass media in the making and unmaking of the New Left*. Berkeley, CA: University of California Press.

Hafsteinsson, S. B., & Bredin, M. (2010). *Indigenous screen cultures in Canada*. Winnipeg: University of Manitoba Press.

Hansen, A. (1993). *The mass media and environmental issues*. New York, NY: St. Martin's Press.

Hansen, A. (2010). *Environment, media and communication*. New York, NY: Routledge.

Hutchins, B., & Lester, L. (2006). Environmental protest and tap-dancing with the media in the information age, *Media, Culture & Society, 28*(3), 433–451.

Hutchins, B., & Lester, L. (2009). Power games: Environmental protest, news media and the Internet. *Media, Culture & Society, 31*(4), 579–595.

Idle No More. (2017). *The vision*. Retrieved from http://www.idlenomore.ca/vision

Keene, A. (2010, September 1). NeverShoutNever and the hipster headdress. *Native appropriations | Representations matter*. Retrieved from http://nativeappropriations.com/2010/09/nevershoutnever-and-the-hipster-headdress.html

Keene, A. (2016). Engaging indigeneity and avoiding appropriation: An interview with Adrienne Keene. *English Journal, 106*(1), 55–57.

Keene, A. (2017). Natives against Redsk*ns. *Native appropriations | Representations matter*. Retrieved from http://nativeappropriations.com/nativesagainstredskins

Kunelius, R., Eide, E., Tegelberg, M., & Yagodin, D. (2016). *Media and global climate knowledge: Climate journalism and the IPCC*. London: Palgrave.

Ledman, A. L. (2012). *Att representera och representeras: Samiska kvinnor i svensk och samisk press 1966–2006* (PhD thesis). Umeå University, Umeå.

Markelin, L., Husband, C., & Moring, T. (2013). Sámi media professionals and the role of language and identity. *Sociolinguistica, 27*(1), 101–115.

McLuhan, M. (1964). *Understanding media: the extensions of man*. New York, NY: McGraw-Hill.

McLuhan, M., & McLuhan, E. (1988). *Laws of media: The new science*. Toronto: University of Toronto Press.

Meadows M., & Avison, S. (2000). Speaking and hearing: Aboriginal newspapers and the public sphere in Canada and Australia. *Canadian Journal of Communication, 25*(3), 347–365.

Melián, V. (2016). The Environment in disguise: Insurgency and digital media in the Southern Cone. In H. Graf (Ed.), *Environment in the age of the internet: Activists, communication, and the digital landscape* (pp. 75–104). Cambridge: Open Book Publishers.

Nisbet, M. (2009). Communicating climate change: Why frames matter for public engagement. *Environment Magazine, 51*(2), 12–23.

Pietikäinen, S. (2003). Indigenous identity in print: Representations of the Sámi in news discourse. *Discourse and Society, 14*(5), 581–609.

Poberezhskaya, M. (2015). Media coverage of climate change in Russia: Governmental bias and climate silence. *Public Understanding of Science, 24*(1), 96–111.

Postman, N. (2000). The humanism of media ecology. *Proceedings of the Media Ecology Association, 1*(1), 10–16.

Pratt, M. L. (2008). *Imperial eyes: Travel writing and transculturation* (2nd ed.). New York, NY: Routledge.

Rodríguez, C., & El Gazi, J. (2007). The poetics of indigenous radio in Columbia. *Media, Culture and Society, 29*(3), 449–468.

Roosvall, A. (2010). Sweden: Between domestication and glocalisation. In E. Eide, R. Kunelius & V. Kumpu (Eds.). *Global climate, local journalisms: A transnational study of how media make sense of climate summits* (pp. 309–324). Bochum: Projekt Verlag.

Roosvall, A., & Tegelberg, M. (2012). Misframing the messenger: Scales of justice, traditional ecological knowledge and media coverage of indigenous peoples and climate change. In E. Eide & R. Kunelius (Eds.), *Media meets climate: The global challenge for journalism* (pp. 297–312). Gothenburg: Nordicom.

Roosvall, A., & Tegelberg, M. (2013). Framing climate change and indigenous peoples: Intermediaries of urgency, spirituality and de-nationalization. *International Communication Gazette, 75*(4), 392–409.

Roosvall, A., & Tegelberg, M. (2015). Media and the geographies of climate justice: Indigenous peoples, nature and the geopolitics of climate change. *TripleC: Communication, Capitalism & Critique, 13*(1), 39–54.

Roosvall, A., & Tegelberg, M. (2016). Natural ecology meets media ecology: Indigenous activists' views on nature and media. In H. Graf (Ed.), *Environment in the age of the internet: Activists, communication, and the digital landscape* (pp. 75–104). Cambridge: Open Book Publishers.

Russell, A. (2005). Myth and the Zapatista movement: Exploring a network identity. *New Media and Society, 7*(4), 559–577.

Russell, A. (2016) *Journalism as activism: Recording media power*. Cambridge: Polity.

Salick, J., & Byg, A. (2007). *Indigenous peoples and climate change*. Oxford: Tyndall Centre for Climate Change Research.

Scolari, C. A. (2013) Media evolution: Emergence, dominance, survival and extinction in the media ecology. *International Journal of Communication, 7*(1), 1418–1441.

Silverstone, R. (2007). *Media and morality: On the rise of the mediapolis*. Cambridge: Polity.

Smith, M., Rainie, L., Himelboim, I., & Shneiderman, B. (2014). Mapping Twitter topic networks: From polarized crowds to community clusters. *Pew Research Centre*. Retrieved from http://www.pewinternet.org/files/2014/02/PIP_Mapping-Twitter-networks_022014.pdf

Stalder, F. (2012). Between democracy and spectacle: The front-end and back-end of the social web. In M. Mandiberg (Ed.), *The social media reader* (pp. 242–256). New York, NY: New York University Press.

Sylvestre, G. (2016). Social network analysis of the 4.6 million #COP21 tweets. *Data Visualization and Social Network Analysis*. Retrieved from https://cartorezo.wordpress.com/2016/02/17/social-network-analysis-of-the-4-6-millions-cop21-tweets/

Tauli-Corpuz, V., & de Chavez, R. (2009). *Guide on climate change and indigenous peoples*. Baguio City: Tebtebba Foundation.

Tegelberg, M. (2010). Canada: The dirty old man of climate politics? In E. Eide, R. Kunelius, & V. Kumpu (Eds.), *Global climate, local journalisms: A transnational study of how media make sense of climate summits* (pp. 275–290). Bochum: Projekt Verlag.

Tracy, C. (2010). A quantum exploration of the news ecosystem. *Explorations in Media Ecology*, 9(3), 193–198.

Tracy, C. (2012). Ecological journalism and the role of perception. *Explorations in Media Ecology*, *11*(2), 131–142.

Treré, E. (2016). The dark side of digital politics: Understanding the algorithmic manufacturing of consent and the hindering of online dissidence. *IDS Bulletin*, *47*(1), 127–138.

Uldham, J., & Askanius, T. (2013). Online civic cultures: Debating climate change activism on YouTube. *International Journal of Communication*, *7*, 1185–1204.

United Nations. (1948). *Universal declaration of human rights*. New York, NY: United Nations.

Williams, H. T. P., McMurray, J. R., Kurtz, T., & Lambert, H. F. (2015). Network analysis reveals open forums and echo chambers in social media discussions of climate change. *Global Environmental Change*, *32*, 126–138.

Wilson, P., & Stewart, M. (Eds.). (2008). *Global indigenous media: Cultures, poetics and politics*. Durham, NC: Duke University Press.

Yagodin, D. (2010). Russia: Listening to the wind—Clientism and climate change. In E. Eide, R. Kunelius, & V. Kumpu (Eds.), *Global climate, local journalisms: A transnational study of how media make sense of climate summits* (pp. 275–290). Bochum: Projekt Verlag.

Young, I. M. (2013). *Responsibility for justice*. Oxford: Oxford University Press.

· 5 ·

ACTIVISM, AGONISM, AGENCY

Indigenous Peoples, Media Witnessing and the Political Game of the Summits

> I think the bigger story is on the outside of the meetings and the relationships and the actions that are taking place with the coalitions, and the indigenous environmental network, and so forth, all of those wonderful actions, and the connections we are making as communities.
>
> —Allison Akootchook Warden, Iñupiaq people (USA, COP21)

In this chapter we discuss the role of *the political* in relation to the role of *politics*, with a point of departure in Chantal Mouffe's (2005, 2013) understanding of the political as the ideological side and politics as the institutionalized side. Most interviewees, many of whom work inside and outside of the UNFCCC summit structure, underline the role of activism for political results and for democracy more broadly. Allison Akootchook Warden stresses that activities outside of the summits in fact constitute *the bigger story*, implying that this is something media should focus on. However, the media focus, to a large extent, has been on what they frame as political games between nation-states negotiating inside the summits (Roosvall, 2010; Roosvall & Tegelberg, 2013). When activities outside the summit are covered, it is often not in ways that (indigenous) activists find relevant or just (Roosvall & Tegelberg, 2015). Here we identify not only a struggle over what the bigger story is, but also a struggle over what broader understandings of the world and of democracy there are from which to "choose" bigger or smaller stories.

We begin by discussing the significance of agonistic democracy and agonistic approaches to communication; that is, the importance of deliberation of a sort which does not necessarily strive towards consensus. While consensus on strict mitigation measures would certainly be useful in climate change discussions, we argue that with issues of (in)justice it is crucial to see beyond consensus, since this has mostly been equivalent to the aims of Western developed nations. Thus, we outline two central remedies for injustice in relation to agonism: (i) representation (i.e. political participation): Who has the right to take part in what discussions? Who has the right to make decisions at the COPs?; and (ii) recognition: Who are recognized as valid speaking-partners? Who are recognized as even having the right to have rights? Specifically in relation to the flaws of the nation-state in terms of stigmatized minorities, deprived of *proper* political membership (see Arendt, 1998, Fraser, 2008, p. 132). Our focus is thereby on the "who" rather than the "what" of justice, heeding an "agonistic cosmopolitics" approach (Hanafin, 2012; Honig, 2009; see also Arendt, 1998).

To subsequently deepen this discussion, we focus on the particular roles of politics and the political at UN climate summits, drawing on examples from media reporting and on statements from the interviewed indigenous activists. In this exploration, we discuss how and where indigenous peoples have a voice in the media (if at all), since in public discourse the lack of ethnic minority voices "in any meaningful sense" (Macnamara, 2016, p. 27, cf. Husband, 2000) constitutes inequality and injustice. We also address a certain branch of activism: *artivism* (Mouffe, 2013). Some of the interviewees are activists and artists, performing art as part of their activism. Allison Akootchook Warden, for instance, is among other things a rapper and performed in this respect during COP21.

In the last section of the chapter, we zoom in on media *witnessing* (Chouliaraki, 2015a; Frosh & Pinchevski, 2009; Peters, 2001) as a typical example of the way indigenous voices are represented in the media. At the COPs, this concerns the witnessing of climate change effects and other human rights offenses. Our analysis of this witnessing is divided into two parts. The first is devoted to what we term "edited media witnessing" (which mainly occurs in mainstream media). The second concentrates on "un-/self-edited media witnessing" (in self- or citizen produced content on social media), which we relate to the intersection of affect, emotion and the political (see Mouffe, 2013; Nussbaum, 2013). This discussion contributes to a growing literature on emotion in media and journalism studies (Beckett & Deuze, 2016; Wahl-Jørgensen, 2016).

Methodologically, the chapter draws on empirical data and interviews conducted in the field.[1] Quotes from interviewees—particularly concerning their views of activism and political processes—are related to theory with the aim of better understanding the interview results and the theories. That is, theories help us understand the interviews, while interviews concurrently help us to better understand theories, particularly when it comes to struggles for justice in an agonistic order. Ultimately, we argue that indigenous activism epitomizes an agonistic cosmopolitics, and helps to clarify how such an approach can be put into practice.

Thus, with a point of departure in the need for intra- and transnational justice in the largely international arena of climate politics, we explore the significance of agonistic democracy and agonistic cosmopolitics in unequal and globalizing times. The roles of indigenous activism at the COPs are examined in relation to this. Furthermore, we examine the relationship between representation in politics and the media. To accomplish these tasks, we address the following questions: In view of the political, in what contexts are indigenous voices heard in media coverage of UN climate summits? How does this relate to other voices and what stories do indigenous peoples tell? More particularly, when indigenous peoples appear as media witnesses, what do they witness about? And, finally, how does this witnessing differ in mainstream versus social and self-produced media narratives?

Agonistic Democracy and Climate Justice: Conflict, Exclusion and Recognition in the Summit Context

When confronting issues of (in)justice, particularly for colonized peoples, it is crucial to look beyond consensus as a goal because it is often equated with the interests of Western developed nations and deeply intertwined with unequal power relations. However, the deliberative democracy model, emanating from Habermas' (1994) understanding of the role of communication in the public sphere, is often drawn upon in media and journalism studies (Strömbäck, 2005). This model underlines the importance of arguments. Ideally, the best arguments win after periods of deliberation (Elster, 1998; Strömbäck, 2005) with the goal or end product being consensus (Gutman & Thompson, 2009). Advocates for radical or agonistic democracy have been critical of the deliberative model, arguing it ignores the inequality of resource distribution between groups (Curran, 2011, p. 81, cf. Macnamara, 2016, p. 14). When discussing issues of justice and the meaning of activism in relation to democracy, it is therefore more pertinent to build on the agonistic democracy model (Mouffe, 2013).

Agonistic democracy is best understood by how it relates to deliberative democracy on one hand and antagonism on the other. Like deliberative approaches it hails arguments, without consensus as its end goal. Conflict is instead seen as essential for keeping democracy on its toes (Roberts-Miller, 2002). *Agonistic* conflict recognizes the presence and necessity of antagonism but assumes the form of agonism in that it is not a struggle between enemies, but a struggle between adversaries (Mouffe, 2013, p. 7). Agonism is therefore often defined by respect (Chambers, 2001; Connolly, 2005, cf. Mouffe, 2013, p. 13). The adversary is an opponent "with whom one shares a common allegiance to the democratic principles of 'liberty and equity for all', while disagreeing about their interpretation" (Mouffe, 2013, p. 13). Adversaries are struggling opponents, with each aiming for their own interpretations of principles to turn into hegemony.

The media can be immensely important in agonistic processes since they can frame events, or grant and deny certain actors voices in media discourses on climate change and justice. This is indeed pertinent regarding indigenous peoples and (climate) politics. An agonistic approach asserts that existing conflicts do not emanate in a "clash of civilizations", which would signify an antagonistic approach (Mouffe, 2013, p. 41). But if agonism is hindered, passions lack a democratic outlet, thereby posing the risk that essentialist identities, such as nationalism and populism, are constructed along the lines of non-negotiable moral values, resulting in violence (Mouffe, 2013, p. 8).

Deliberative and antagonistic approaches can also be combined in calls for consensus (deliberative approach), with antagonistic tendencies kicking in to force a consensus if there is opposition. In agonistic politics, confrontation is instead realized in contexts regulated by democratic organization accepted by the adversaries (Mouffe, 2013, p. 9). This gives reason to question the UN climate summits, where indigenous peoples do not accept the procedures and struggle to have their interpretations turned into hegemony, knowing full well that they have a long way to go. Jannie Staffansson describes her experiences of being excluded from meetings where, as an Observer,[2] she should have been allowed to participate:

> As observers we do [have access] part of the time. But in Bonn [at the pre-Paris conference], the COP decided to shut the door for observers when the negotiation on the ADP text was happening. So we were thrown out of the room … We need to know what state says what and what's happening, but it appears they're sending us

> out again. So that's unfortunate and very frustrating, disturbing ... There are millions of indigenous peoples that are not represented during these negotiations and now we are also blind. We do not know which country said what, and we can't verify it. (Jannie Staffansson, Sami, Samiland, Sweden, COP21)

The "blindness" that results from this exclusion hinders indigenous activists and politicians from formulating arguments that would contribute to the process of agonistic democracy. Instead, there is a lack of knowledge of what arguments the adversaries have formulated. In view of Arendt's metaphor of the public realm as an agonistic "space of appearance", with the possibility of not only seeing but also being seen (Chouliaraki, 2013, p. 23), it is notable that invisibility comes with not being heard. Connecting to this aspect, Lene Kielsen Holm tells a similar story of exclusion:

> The mornings are starting with the regional caucuses ... this morning, I was appointed to ... try to meet with the nations that are representing us. For instance, for me, it's Denmark and the Greenland government. I went there, but I couldn't get ... inside because it was closed, only to the states and the delegates that are within the states. And I'm not in that delegation. So this is just to stand outside ... and try to look at who there is and lobby those that you can catch there. (Lene Kielsen Holm, Inuit, Greenland, COP21)

Holm points to the importance of not only seeing what is going on, as Staffansson underlined, but also having a voice and being listened to. She continues to strive for this by lobbying outside of the meetings after being excluded from participating inside. Since democracy builds on *vox populi*, the voice of the people (Fishkin, 1995; cf. Macnamara, 2016, p. 26), it concerns not only representation, but also *articulation* (Chandler, 2009, p. 7; Chouliaraki, 2013, p. 23). Hence, it is important to remind ourselves that Fraser's justice dimension of representation (political participation) includes the demand for possibilities of articulation, which largely take place in media and politics today. While voice can be seen as the "implicitly linked practices of speaking and listening" (Couldry, 2009, p. 580), limited attention has been paid to the listening aspect (Macnamara, 2016, p. 29; see also Russell, 2016). We shall return to listening later in this chapter, relating it to witnessing on diverse media platforms.

Regarding degrees of participation it is noteworthy that Holm is a Delegate with the Inuit Circumpolar Council (ICC), not the Danish or Greenlandic delegations. Staffansson, of the Sami people in Sweden, represents the Sami Council as an Observer. This grants her access—at least until thrown out—but not the right to take part in decision-making. Another interviewee, who chose to remain anonymous regarding this specific comment, commented on the lower status of being an observer: "It is window dressing". The same interviewee also noted that Norway, the only Nordic country that had given accreditation to Sami representatives, had just degraded the Norwegian Sami, meaning they could no longer participate in the group that negotiates. The different roles for Sami in the Swedish and Norwegian contexts highlight differences in democratic conditions for indigenous groups, which vary even more around the world:

> … for [indigenous groups from] Western countries, for small countries … which closely work with their governments it is much easier. For [indigenous groups from] other countries like Russia, like China, it's much more difficult to [have] influence on the governments. So … there … [are] different levels of negotiations and understanding. (Dmitry Berezhkov, Itelmen people, Russia, COP21)

Democracy is threatened when non-hegemonic groups are made blind to arguments posed by hegemonic groups, as well as when they are muted, denied a voice and ignored at summits and in the media.

The media are arenas for diverse levels of access and participation. There is the level of *presence* in media coverage, in terms of being included in stories but without a voice. There is the level of *dialogue*, where actors are allowed to voice their opinions. Finally, there is the level of *political deliberation*, where actors are included as political beings who are capable of voicing their political opinions (Roosvall & Tegelberg, 2012).[3] To demonstrate this, let us zoom in on the Arctic region, since it is often used to exemplify climate change in public discourse (Christensen, Nilsson, & Wormbs, 2013), and its indigenous peoples who are often chosen to "speak for the climate" (Bjørst, 2012). In Swedish and Canadian reporting during COP15 (2009) and COP17 (2011), 11 articles mentioned indigenous peoples from a total of 598 examined articles. 10 of these articles focused on Arctic indigenous peoples.[4] In the elite Swedish newspaper (Dagens Nyheter), Arctic indigenous peoples only appeared at the level of *presence*, while they appeared at the level of *dialogue* in the popular press in both countries. The level of *political deliberation* was only touched upon vaguely in a small number of the articles featuring Arctic indigenous peoples. In the 11 articles, all levels constitute forms of *indirect participation*, which can be contrasted with the forms of *direct participation* that take place in social and self-produced media, where journalists do not frame your voice (see Roosvall, 2016a).[5]

The blinding and muffling of indigenous peoples at climate summits, and a corresponding lack of voice in media, reminds us of Arendt's (1951) contention that just because rights are proclaimed, they are not necessarily politically secured. Therefore it is crucial to point to the limitations of institutionalized rights discourse by speaking of the "right to have rights" (Hanafin, 2012, p. 327). It is not enough that democratic participation in the public sphere is "inclusive and reflexive, it must also be effective" (Nash, 2014, p. 1). As Hanafin (2012, pp. 326–327) writes, it is a trap to believe "that the existence of a bill of rights automatically leads to freedom for all. While striving for such a goal is laudable, it should not be confused with the achievement of freedom as such". What is evoked by this approach, and the accounts of being blinded and muffled at the COPs, is injustice in the form of political *misrepresentation* (Fraser, 2008). There is thus a need for a reframing; not just an affirmative politics of reframing (concerning certain cases) but a transformative politics of reframing that concerns all indigenous groups and the UN decision-making process.[6]

When it comes to framing in the agreements produced at the COPs, some activists emphasized the need to be *recognized*, thereby evoking *misrecognition*; yet another dimension of (in)justice:

> Activism here in the COP is—we as activists push more the politician to *recognize* our demand as indigenous peoples, and also to *recognize* us as marginalized people … So we are expecting, as youth and as indigenous peoples, to have a document wherein … to be *recognized*, and also to have support from different governments … to support us in our struggle—to fight for our human rights. (Justeene Dammay, Tuwali people, Philippines, COP21, emphasis added)

To be recognized is to be treated justly (Fraser, 2008). Dammay stresses the need for this, using the term recognize repeatedly. Moreover, she mentions human rights, which again underlines that climate change is viewed as an issue of justice. It has to do not only with recognition of group identity, but also of group status. Her account furthermore shows that achieving this entails a struggle or conflict, thus evoking agonistic approaches.

Marie-Josee Artist also discusses the importance of recognition and how the agreement frames climate change in relation to indigenous peoples globally:

> I think that because we don't have our lands *recognized*, legally *recognized*, for us it is very important that there is something in the agreement, in the binding agreement, about the rights of indigenous peoples so that our rights are secured there and then

> we can fight on national level to really have it. (Marie-Josee Artist, Kalina people, Suriname, COP21, emphasis added)

Like Dammay, Artist stresses recognition and underlines that this is a fight. She emphasizes that it is hugely important to come out of this struggle with your rights recognized in a binding agreement. Both interviewees thereby connect to the constitution of collective identities, which is a crucial ingredient in treaties as well as in media output. Arendt (1998, p. 52) states: "the world, like every in-between, relates and separates men at the same time". A crucial question is thus who, when, where, how and what is being related and separated, and how this is performed by the media. Is it an agonistic or an antagonistic process? When it comes to indigenous peoples, it is relevant to distinguish them from others in terms of *historical experiences* (i.e. pre-colonial and colonial pasts), their *collective identities*, as well as in regards to present and expected acute experiences of climate change on their (occupied) lands (postcolonial present and future). At the same time, indigenous groups can be included in a collective that encompasses *all of humanity* regarding norms for a tangible shared equal status, and with regards to common transnational challenges posed by global climate change. Determining whether this happens in politics and the media is a crucial task for research relating to these areas. We tackle this further specifically in the section on media witnessing (below) and in Chapter 6.

The examples of misrecognition, blinding and muting illustrate the need for new ways of approaching human rights since the existing models are not effective. Hanafin (2012, p. 327, cf. Cavarero, 2000, 2002) aims to rethink the relationship between cosmopolitanism and human rights, with a starting point in Hanna Arendt's (1951, p. 296) ideas on "the right to have rights." An *agonistic* cosmopolitics of human rights sees individuals as *active participants in political affairs* rather than legal subjects (Hanafin, 2012, p. 327, cf. Honig, 2009, p. 130). Hence, the political participation (representation) dimension of injustice is addressed in the very rationale for this thinking. Balibar (2002, p. 167) accordingly reformulates "the right to have rights" into "the right to politics" (see also Hanafin, 2012, p. 328). The right to have rights implies that people belong to a *political* space (Birmingham, 2007, p. 771; Hanafin, 2012, p. 328). Hanafin (2012, p. 331) sees this as a global space of interaction, not demarcated by state borders, where there is a "… thinking of politics beyond the bureaucratic rights-giving or rights-depriving state, thereby connecting to the rethinking of justice in a globalizing age (Fraser, 2008), and to Young's (2013) social connection model of responsibility.

Development of the right to have rights or the right to politics in global spaces is particularly pertinent for indigenous peoples since in existing systems "paradoxically, we need rights because we cannot trust the political communities to which we belong to treat us with dignity and respect; however we depend for our rights upon those very same political communities" (Honig, 2009, p. 117). Furthermore, according to Arendt, to think politically is to be able to see things from multiple perspectives (see Mouffe, 2013, p. 10). This combined with Arendt's (1998) conceptualization of *the subject* of political philosophy as a unique embodied self—as a "who" instead of a more abstract "what" (see Hanafin, 2012, p. 330)—makes her perspective specifically fruitful for understanding indigenous rights, and indigenous rights claims. The recognition of the subject as a person as well as the recognition of the existence of multiple perspectives should be productive for promoting indigenous rights. Cavarero (2002, p. 99, cf. Hanafin, 2012, p. 330) says the "who" "… has a face, a name and a story." It is the stories of these subjects with faces and names, that Allison Akootchook Warden calls for the media to focus on more (see beginning of this chapter); namely, outside of official negotiations at the COPs, where multiple perspectives can be found. India Logan-Riley also evokes the necessity of multiple perspectives and a focus on the "who" of justice in media reporting on climate change:

Interviewer:	How could media coverage, mainstream or alternative, be better then, do you think? I mean coverage of climate change.
India Logan-Riley:	I think it could be contextualised better, rather than just a single episode … give it a more comprehensive frame of where we've been and where we're going rather than a few quotes and sort of a half page article basically. I think also it could give more diverse views [than] … the stereotype view. So it would be good to have Māori perspective, Pacific Island perspectives … Give a human face but give it context. (India Logan-Riley, Māori, New Zealand, COP21)

Logan-Riley connects here to the need for the "who" to have a face, as well as the recognition of multiple perspectives, which both characterize an agonistic cosmopolitics. It is notable that the recognition of multiple perspectives does not mean there are absolute essential differences between people. Aiming to overcome dichotomous distinctions Cavarero (2000; see also Hanafin, 2012) builds on Arendt by talking about the "who" not as the Other but rather as "another."[7] She writes that this "another" is a "you" that must be brought into strong focus in relation to the "I" in contemporary political discourse, especially since "[t]he you is ignored by the individualistic doctrines, which are too concerned with praising the rights of the I" (Cavarero, 2000, p. 90, cf. Hanafin, 2012, p. 331). We can assume that such a "you" is what Marie-Josee Artist wants included in the final agreement at COP21 (see above), which is otherwise written from the point of departure of a state-determined I. This state-determined I, in fact, permeates institutionalized "politics" inside the COPs and the bulk of the media reporting surrounding these events. This stands in stark contrast to the more ideological character—"the political"—of the activism that goes on "outside". Thus we now elaborate further on this distinction between politics and the political.

Politics Versus the Political: What Is the Story and Who Tells It? Varieties of Voice and Agency

A central distinction for understanding indigenous activism during the climate summits is Mouffe's (2005, p. 8) differentiation between "politics" and "the political". "Politics" refers to conventional politics: "… [the] set of practices and institutions through which an order is created" (Mouffe, 2005, p. 9). Politics orders human life and coexistence in relation to the *conflictuality* created by "the political." Thus, "the political" is connected to agonism. Herein lies the recognition of social division and the legitimation of conflict (Mouffe, 2005, p. 119). "The political" foregrounds the existence of a plurality of interests and demands that must be considered, notwithstanding that all needs can never be fully reconciled. In fact, if social division cannot be expressed in a *political* way, passionate mobilization for democratic goals can be replaced by antagonisms that will threaten democratic institutions (Mouffe, 2005, p. 120). The political can be seen as an open space of *communication* and interaction between individuals (Arendt, 1998; Cavarero, 2005; cf. Hanafin, 2012, p. 331). Thus the roles of mediated communication in relation to the political become essential for democracy (see also Fenton, 2016). In the next two sections we discuss expressions of "politics" and "the political" in relation to the climate summits. In both cases, we relate these expression to media output and to interview material.

Politics: The Political Game Frame, the (Inter)national Scale and Domestication

"Proper political questions always involve decisions that require making a choice between conflicting alternatives", writes Mouffe (2013, pp. 3–4). This can be imagined as a pertinent means of describing how the COPs work, with nation-states negotiating and promoting conflicting measures. But can these be considered alternatives in a proper sense? Often summit politics are framed by the media as a "political game" (Roosvall, 2010; Roosvall & Tegelberg, 2015).[8] While this may suggest conflict, it is not necessarily portrayed as a conflict between ideologies. On the contrary, these conflicting nation-states mainly promote the same neoliberal ideology, often working to protect themselves from having to yield the privilege they enjoy. The issue at stake, climate change, is often lost in this game frame and is thus symbolically annihilated (see Tuchman, 1978) by such media reports. According to Strömbäck (2005, p. 342), failure to inform the public about the issue at stake is one of three main reasons why journalism should refrain from framing politics as a strategic game. The other two reasons are that such framings position ordinary people as spectators rather than as political actors, and that this fosters cynicism and distrust (see also Cappella & Jamieson, 1997; Strömbäck, 2004).

In Swedish coverage of COP15 (Copenhagen, 2009), the "political game frame" was the most commonly used among a number of news frames (81 of 158 articles).[9] The "issue frame" which focused on long-term perspectives and on climate change as an issue rather than as a political game, was used less (43 of 158 articles) (Roosvall, 2010, p. 314; see also Tegelberg, 2010).[10] When indigenous peoples appeared in Swedish and Canadian coverage of COP15, it was mainly within this less prominent frame (Roosvall & Tegelberg, 2013, p. 405). One interviewee related to the disconnection of these two main frames while discussing climate change, indigenous peoples and media coverage of the COPs:

> [The] media manipulates the subject of climate change … I would like to see that it's not manipulated. Especially when it comes to the questions of indigenous people … it's … necessary to observe the true story, the real situation, and real data … people have to know what's going on. *And that is why it's necessary to listen [to] the truth about the climate change … not only talk about the political issue.* (Victoria Sharakhmatova, Itelmen people, Russia, COP21, emphasis added)

This disconnection is also related to the evocation of different geographical scales: the "issue frame" largely expressed connections between the local and the global, while the "political game frame" was generally connected to the nation-state, the international scale and domestication (Roosvall, 2010, p. 321).

Similar patterns of national domestication were evident in media coverage from around the world during several climate summits (see Eide &

Kunelius, 2012, p. 23; Roosvall, 2010, p. 312). These evocations of nation-state determined frames constitute misframing in Fraser's terms (2008), that is, the imposition of nation-state frames on transnational issues, like climate change, that effect transnational and intranational groups, such as indigenous peoples, in non-state-determined regions. This imposition constitutes the worst form of injustice that can occur in a globalizing world (Fraser, 2008). Like the Swedish material, the examined Canadian articles from COP15 mainly focused on the summit. The four stories that included indigenous peoples framed climate change as a domestic and/or global *issue* rather than as a political game unfolding between nation-states in Copenhagen. The indigenous peoples featured in this coverage appear to be struggling to connect these separate frames; struggling to make connections between nature and politics, and break with nation-state determined frames (Roosvall & Tegelberg, 2013). In a *Toronto Star* article, for example, Inuit activist and political representative, Sheila Watt-Cloutier is quoted as follows:

> It isn't just about ice and snow. … It's about people, and it's not just about the furry animals that the people want to focus on saving. It's about the sustainability of communities in the Arctic and other places that really rely upon the well-being of their environment to be able to continue to practise an ancient culture that is extremely respectful to all things around us. (Woods, 2009)

Watt-Cloutier stresses that nature and humans are connected. Moreover, her emphasis on sustainability makes this a political issue that is not limited to the nation-state but rather one that is connected to *communities in the Arctic*, thus evoking a transnational scale.

The *national* political system however features prominently not only in terms of framing, but also in terms of the voices represented in Swedish summit journalism (Roosvall, 2010), as well as in several countries around the world. Figure 5.1 visualizes the voice data collected from newspapers in 11 countries. The national political system includes any voices operating on the national scale, whether they are domestic or foreign representatives of a national political system. Voices representing what are here termed "transnational politics" generally originate from groups or organizations that are made up of conglomerations of nation-states, like the EU and the UN. Thus, they are not transnational in the same sense as we defined it in Chapter 1. Instead we argue that these voices largely adhere to the logic of the nation-state. Painter (2010, p. 4) adds that even when science was reported during COP15, "… it was representatives of international and national bodies (including governments) who were by far the most quoted." The truly transnational voices in Figure 5.1 are likely hidden in the "civil society" category which consists of grassroots organizations, NGOs and social movements that work along several scales, including the transnational.

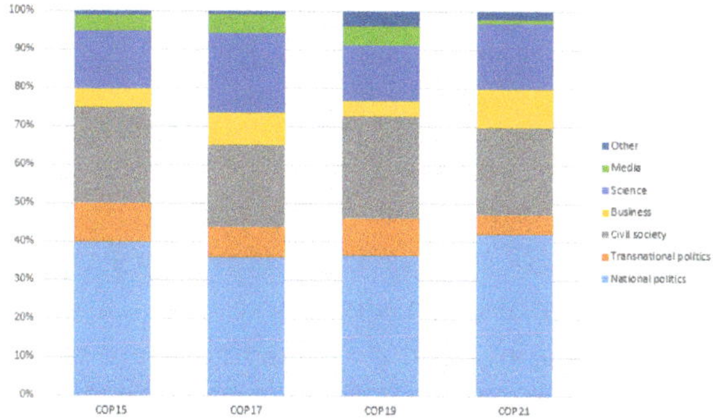

Figure 5.1. Principal groups of voices quoted in reporting on four UN climate summits.[11]

If they appear at all, indigenous peoples and indigenous activists would most likely appear among the civil society voices. The civil society category is also where "the political" is most likely to be expressed. However, as established in Chapter 4, indigenous voices are rarely included in mainstream media coverage of the COPs.

We asked Pirita Näkkäläjärvi, head of Yle Sápmi (Sami broadcasting in Finland, part of the Finnish national broadcasting): How do you think your reporting and media outreach activities during COP21 differed from reporting done by the mainstream media? She replied:

> The difference is clear: the focus of our reporting was always on the Sami and indigenous perspectives, and on indigenous rights. We interviewed representatives of Sami people (Sami Parliaments, Sami council and Sami associations), Sami activists from Finland, Norway and Sweden as well as … statements by Victoria Tauli-Corpuz, the UN Special Rapporteur on indigenous issues, or by the IIPFCC (International Indigenous Peoples' Forum on Climate Change) … Looking at a couple of examples by the Finnish Yle, their focus was on the overall status and content of the Paris climate agreement. The Finnish Yle interviewed representatives of the Finnish state, quotes officials of the COP21 and published critiques by Finnish activists. (Pirita Näkkäläjärvi, Sami, Samiland, Finland, Head of Yle Sápmi, e-mail interview 2017)

Näkkäläjärvi thereby points to a vast difference between what voices are heard in mainstream and indigenous media. Moreover, by integrating voices and perspectives indigenous media are able to encompass a truly transnational dimension that seems difficult for mainstream media to capture. Näkkäläjärvi also emphasized that: "Every Sami person is an activist of a kind. No Sami professional can avoid accusations of being too political, because Sami life is so inherently political." In light of these sentiments, we shall now zoom in on the political and, more specifically, on indigenous activism.

The Political: Activism, Its Indispensability and Its Suppression

As part of an agonistic cosmopolitics approach, "the political" could help us "restore to rights some of their pre-institutionalized radicality" (Hanafin, 2012, p. 327). This constitutes an alternative or a complement to institutionalized politics, by focusing on the very right to have rights. Activism has a crucial role to play in this regard, as stressed by many indigenous activists:

> Activism is vital for all politics. Like if activists didn't speak up, women wouldn't have voting rights for example, and that's human rights and that's vital for all politics. So people speaking up and communities speaking up, it's vital and … Yeah, it means everything. (Jannie Staffansson, Sami, Samiland, Sweden, COP21)

> … if you just sit inside these halls … if you expect that to change things … you're dreaming … you've got to be everywhere … if governments don't move then people have to move them, and in order for that to happen then we as activists working within the framework of the climate change convention must organise inside [the summit structure], but it's just as important to organise outside—the social movements. (Teanu Tuiono, Māori, Cook Islands and New Zealand, COP21)

The crucial role of activism, and "the political" more generally, has also been identified in media research. Recent literature has concentrated on media power and counter power (Cammaerts, Mattoni, & McCurdy, 2013); on media produced by protest groups and protest tactics (Cable, 2016); on the role of social media as organizing mechanisms (Segerberg & Bennett, 2011); on reaching broader publics *and* mobilising for confrontational direct action within activist circles (Askanius & Uldham, 2011); on the roles of media practices and technologies in relation to economic crisis over time (Kaun, 2016); and on journalism *as* activism (Russell, 2016). *Indigenous* (environmental) activism can relate to all these aspects, but must at the same time more particulalry be understood in relation to colonialism and struggles over treaty rights (Clark, 2002). It has moreover been recognized as specifically transnational (Plaut, 2012). Keck and Sikkink (1998) discuss how indigenous peoples—among other activist groups—form action networks beyond borders. Lightfoot (2016, p. 4), in turn, terms indigenous politics *global*, identifying the indigenous rights movement as a "*transformational norm vector*, pointing the way toward some alternative ways of doing global politics and new imaginings of global order that can potentially come to exist" (emphasis in original). This resonates with both an agonistic cosmopolitics of human rights (see above) and with the tackling of misframing through a transformational politics of framing (Fraser, 2008), questioning not only the nation-state frame in certain instances, but doing it across the board.

Political perspectives have however been downplayed in mainstream media reporting on indigenous peoples and climate change. On the few occasions that indigenous peoples appear in climate summit journalism the focus has been on culture and spirituality (Roosvall & Tegelberg, 2013). In fact, spiritual and political discourses seem to be excluding each other in articles and reportage concerning how climate change affects indigenous peoples in Swedish and Canadian media. This mirrors the broader tendency in Western public discourse not to mix politics and religion (Christi, 2001; Roosvall, 2011), and the fear of Native American spiritual practices that accompanied xenophobia when Europeans first arrived in the Americas (LaDuke, 2005, p. 11). While spiritual perspectives on human-nature relationships nowadays can be portrayed as likable, they may at the same time be represented as wrong. We return to this in the section on media witnessing.

At COP17 in Durban, we asked a non-indigenous representative of a group working to protect national heritage sites how media coverage of indigenous peoples and climate change could be improved. His response was that "I think the wearing of your indigenous and your cultural clothing actually says much more than talking about it" (Anonymous, International National Trusts Organization, COP17). This interviewee suggests that in order for media coverage to improve it is indigenous peoples rather than the media that need to change. The interviewee simultaneously implies that indigenous peoples' voices—who could be speaking about the politics and the political of climate change—are not as important as the appearance of indigenous peoples as cultural beings. Regrettably this interviewee has a point; not in that indigenous peoples should change rather than the media, but in that while cultural traits have generated some attention, political perspectives have been harder to get into the mainstream media.[12] However, this may be beginning to change as reporting on indigenous led resistance to the Dakota Access Pipeline (the NoDAPL movement) in the United States has indicated. As we shall see, there are also tendencies towards the political in Swedish summit journalism from 2015.

There are at the same time different forms of representation in terms of relationships to the who of justice in different types of media. Indigenous mediated self-representations, and representations produced by allied organizations, seem more likely to address a "you" and "another" rather than "the other", which appears in mainstream media accounts (Pietikäinen, 2003), even when they are empathetic (see Roosvall & Tegelberg, 2012). For example, during COP21 Amazon Watch carried the headline "Humanity—Together—Is in Danger of Extinction. Are We Really Understanding This?" (Amazonwatch.org, 2015a). This alludes to a Common Human Identity where people must come together through addressing a "you" or rather a "we". The headline and story illustrate Arendt's contention that the polis is not "the city-state in its physical location; it is the organization of people as it arises out of acting and speaking together, and its true space lies between people living together for this purpose, no matter where they happen to be" (Arendt, 1998, p. 198, in Hanafin, 2012, p. 331). This is epitomized by the acting and speaking of indigenous peoples from around the world inside and outside of the official political negotiations at COP summits, as exemplified by Allison Akootchook Warden (above) when she underlines "the relationships and the actions that are taking place … the connections we are making as communities". Such a "being together" is, as Hanafin (2012, p. 331) writes, "not in the name of an object such as a nation or a territory, but rather the reverse, a being together outside and beyond these imposed hegemonic notions of political organizations."

There are also diverse forms of representations in terms of voice in diverse media. The first entry in the aforementioned Amazon Watch article includes voice, problematizes it and addresses listening:

> *We* have to come to France to have my *voice heard* and all of the *voices* of Sarayaku represented—we need *the world* to *listen* to our *voices*, especially as women who care for future generations. *We* live in the heart of this *shared planet*—we are *defending it not just for ourselves*, *the world needs to hear this*. Ena Santi, Sarayaku leader (Amazonwatch.org, 2015a, emphasis added)

The next entry reads:

> "Let's leave our *old perceptions and ideas*, to be *reborn* and boost a collective *transformation*!" Patricia Gualinga, the Kichwa Director of International Relations from Sarayaku told a packed house at the Rights for the Earth conference at the UN COP21 climate summit in Paris. "Now is the time to stop these destructive ideas in

> our ancestral territories and around the world. We as indigenous peoples have the great opportunity to *bring our vision*—and a *clear proposal*—that could call on a *transformation* for all of *humanity*. (Amazonwatch.org, 2015a, emphasis added)

The Amazon Watch article not only problematizes the politics of voice, but also explicitly reports on alternative political activities in Paris; thereby becoming an arena for the expression of a transformative politics of framing (Fraser, 2008), and of the creation of alternative worlds as detailed in agonstic cosmopolitics theory (Hanafin, 2012). One such alternative world is described as follows:

> Kawsak Sacha—the "Living Forest"—is Sarayaku's proposal for living in harmony with the natural world, stemming from the knowledge of Indigenous peoples who have inhabited the Amazon for millennia. While much of the western world views nature as a source of raw materials for human use, Kawsak Sacha recognizes that the forest as a whole is made up entirely of living beings and the connected communicative network that they have with each other. According to Sarayaku "the Living Forest economic system is an ecological web; the natural world is also a social world. (Amazonwatch.org, 2015a)

AmazonWatch worked closely with the Kichwa community from Sarayaku, Ecuador during COP21. One prominent event was a sunrise ceremony to mark the arrival of "the canoe of life" to Paris on December 8. An Amazon Watch press release for the event read:

> "Some 500 years ago the conquistadores came to our lands on boats, bringing devastation and death," said Felix Santi, President of Sarayaku. "Our canoe is a symbol of life and peace from the heart of the Amazon to show the world that our forests are still living and to tell those negotiating a climate treaty that *indigenous rights and voices must be a fundamental part of any agreement* that hopes to address climate change." (Amazonwatch.org, 2015b, emphasis added)

This sunrise ceremony, and a flotilla action held the day before, which included several canoes, can be seen as expressions of kayaktivism (see Bagelman, 2016). Both actions demand that indigenous rights be included in the COP21 agreement (see also the emphasis in the Amazon Watch press release above, where voices are highlighted). The flotilla event was widely reported by international media outlets like *The Guardian*. Allison Akootchook Warden connects to kayaktivism when describing the power of activism:

> I think the change that comes, comes from the grassroots, or from the push from the grassroots people. For instance, I was involved with the decision for Shell to pull out of the Arctic Ocean, and that decision … came from many, many different factors

> and many different people who were working on that in levels of policy and also through grassroots organizing ... the kayaktivists and the activities that happened in Seattle ... and also Portland, and other places around the world, in protest of Shell. I have seen firsthand that those kind of grassroots activities and people rising up result in the companies responding and doing what is right. And we saw that directly with Shell pulling out of the Arctic Ocean, which was good. (Allison Akootchook Warden, Iñupiaq people, USA, COP21)

Another Amazon Watch press release for the Paris flotilla event places emphasis on visuals: "Indigenous flotilla of several canoes and at least 25 kayaks *adorned with Indigenous art work* representing the different Indigenous cultures participating in the event" (Amazonwatch.org, 2015c, emphasis added). The mentioning of art work implies that this event can also be seen as "artivism" (Mouffe, 2013, p. 99). The potential of artivism as a strong force in agonism and in the political lies in its ability to produce new subjectivities and new worlds, to transform common sense through counter-hegemonic moves (Mouffe, 2013, p. 87, 90). Several activists that we interviewed can be described as artivists. Allison Akootchook Warden is one of them:

> I'm an artist ... there's many of us who are artists/activists ... my heart and my work is grounded in art ... Art is the medium in how I respond to these things ... I'm a rapper. So I do music, and I rap as the animals. And then, I'm doing a show specifically around climate change and the impacts that my people will be feeling, not just with the climate, but with increased shipping lanes, and also, the push to drill for more resources. So I'm having an installation performance art at the museum in Anchorage. (Allison Akootchook Warden, Iñupiaq people, USA, COP21)

Atsynga Olga Letykai Csonka likewise made several appearances as a throat singer in the Generation Space area at COP21. Jordan Konek, a member of the Inuit Youth Delegation (interviewed at COP17), described some of the videos he produced for YouTube and the Nanisiniq Arviat History Project website (http://nanisiniq.tumblr.com/). Outside of the climate summit context we found that whatlocalpeople.se, a website designed to support the struggle against mining activities in Gállok (Kallak) in Sapmi, Northern Sweden,[13] included several expressions of artivism including videos of the performances. Other videos and a spoken word performance connected to the struggle were posted on the YouTube channel "whatlocalpeople" (Cocq, 2016).

At COP21, daily activities at the Indigenous Peoples' Pavilion featured intersections of political and artistic expression. This included singing, dancing, film screenings where culture and the political were intertwined.

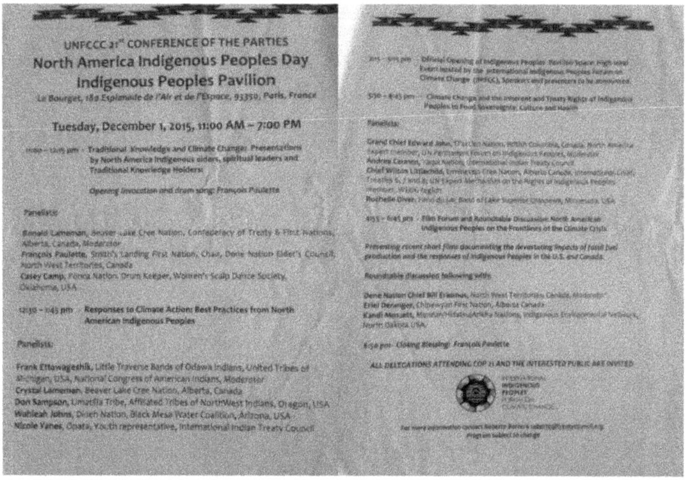

Figure 5.2. Program for North America Indigenous Peoples Day at the Indigenous Peoples' Pavilion.

The agenda for North America Day included drumming and a film screening together with discussions on climate action, treaty rights and traditional knowledge (see Figure 5.2). The official opening of the pavilion also took place on North America day. This event included throat singing and drumming from different parts of the world, along with political remarks. In addition, the prayer ceremony that we described in Chapter 1 included several indigenous activists and spoken word poets who had won local contests that enabled them to travel to Paris; and around the city, several arts exhibitions took place that focused on indigenous peoples. For example, the Point Ephemere cultural centre featured the photo exhibition "If Not Us Then Who?" along with a series of topical film screenings. In the entrance to the exhibition a text with the headline "Indigenous Peoples and Local Community Demands" was followed by four points: "Recognition to Land, Territories and Resources", "Consent", "Zero Violence" and "Funding". The photographs depicted indigenous peoples from around the world, including the widow of an indigenous activist who was murdered defending his territory and people fighting to preserve forests, among others. Advertisements in the subway drew attention to the website "If Not Us Then Who?" (http://ifnotusthenwho.me/), using slogans like "Stop climate change. Land rights for indigenous peoples", "Indigenous peoples protect forests. Respect their rights." The ads also featured the hashtag #paddletoparis. "L'Amazonie filmée par se peuples" (the Amazon filmed by its peoples) took place at the Quai Branley Museum. In Mouffe's (2013, p. 100) view, museums could be transformed into agonistic public spaces, contesting hegemony. Mouffe specifies that hegemony concerns consumer society. In the case of climate change and the UN summits, hegemony simultaneously concerns the geographical scales of the nation-state and international relations; ways of viewing nature, as well as the conceptualization of rights, illuminated by the counter-hegemonic activities of indigenous artivists and activists.

In our interview with Pirita Näkkäläjärvi, we asked what stories were the most important ones Yle Sápmi reported during COP21. She answered that:

> Our most important stories were the ones covering Sami activism ahead of COP21: the We Speak Earth yoik sang by our TV anchor in the TV news broadcast, Jenni Laiti starting the Run For Your Life relay (with our reporter there in Kiruna working on a climate special) and the story from the London demonstrations just before COP21. (Pirita Näkkäläjärvi, Sami, Samiland, Finland, Head of Yle Sápmi, e-mail interview 2017)

Yoik is a traditional Sami way of singing. It "creates an emotional connection between humans, animals and nature. [People] do not yoik *about* something the way you do when you sing. [People] *yoik something*. Thus becoming a part of that which is yoiked" (samer.se, n.d., emphasis added). The inclusion of this We Speak Earth yoik in the Yle Sápmi's broadcast constitutes an example of artivism—connecting art and activism—that merges with journalism as well, as this quote from Pirita Näkkäläjärvi shows:

> In October 2015, prior to COP21, our news anchor Ailu Valle (also known as a rapper) yoiked in a TV news broadcast (link below)! More specifically, he yoiked the We Speak Earth yoik made by activist Sara Marielle Beaska Gaup (Internet challenge October 2015: https://www.youtube.com/watch?v=H2LhBAi-Q8I) … the news story with the video to Ailu's yoiking was shared a lot in social media. (Pirita Näkkäläjärvi, Sami, Samiland, Finland, Head of Yle Sápmi, e-mail interview 2017)[14]

The Yle Sápmi news anchor's yoiking thus expresses not just journalism as activism (Russell, 2016) but *journalism as artivism*.

In describing an agonistic cosmopolitics, Hanafin (2012, p. 337) writes that "When groups of individuals come together to refuse the symbolic disembodiment which the social contract offers them, they expose the limits of law, its internal paradoxes, and the impersonality of the person with rights". This resembles our description of indigenous peoples actions at and around the climate summits, and thereby shows the crucial roles these groups can play to enhance understanding of the issues of rights, justice and responsibility. The radicalness of this approach lies in a "refusal to accept certain modes of thinking"

(Hanafin, 2012, p. 337). Mouffe (2013, p. 97) underlines the specific potential of art in such endeavours: "As Dewey pointed out, works of art allow us, through imagination and the *emotions* they evoke, to participate in new experiences and to establish forms of relationships that are different from the ones we are used to" (emphasis added). In the next section, we discuss media, emotion and affect further by zooming in on witnessing in various media channels.

Varieties of Media Witnessing, Varieties of Voice and Affect

In this section we discuss media witnessing (Frosh & Pinchevski, 2009), making a distinction between what we term *edited media witnessing* and *un- or self-edited media witnessing*. Edited media witnessing is carried out in professional journalism, often in mainstream media, where the witnessing is edited by the journalists constructing the stories. Here we draw on examples of indigenous peoples witnessing in Swedish and Canadian media coverage from COP15 (Copenhagen, 2009) and COP21 (Paris, 2015). Un- or self-edited media witnessing occurs on platforms such as YouTube and Facebook, where there is no editing or the editing is conducted either by members of groups or groups collaborating with them. Here we draw on examples from YouTube, Facebook and online reporting produced by NGOs during COP21. These two types of media witnessing differ in terms of the direct or indirect participation of those doing the witnessing. They also differ based on levels of presence and agency; what the witnessing is about; emotion and affect; geographical scales and in terms of whether the witnesses speak in the name of a we or as a we (see Table 5.1). In the ensuing discussion, we exemplify and elaborate on how indigenous peoples appear in edited media witnessing and, subsequently, in un-/self-edited media witnessing.

Table 5.1. Edited versus Un-/self-edited media witnessing.

	Edited media witnessing	Un- or self-edited media witnessing
Participation	Indirect (framed by journalists).	Direct. Potentially indirect at times (unframed or framed by themselves).
Level of presence	Mostly at level of dialogue.	Mostly at level of political deliberation.

(*Continued*)

Table 5.1. (*Continued*)

	Edited media witnessing	Un- or self-edited media witnessing
Agency	Reactions—to journalists' questions, within a journalistic framing. Conditional agency.	Actions. Unconditional agency. Political agency.
Speaking in the name of/as	Speaking in the name of a we.	Speaking as a we.
Witnessing about	*Weather/climate*, plus spiritual, cultural and to some extent political aspects (the latter more pronounced in 2015).	*Political aspects* and weather/climate, sometimes combining the political and the spiritual.
Expression of emotion, affect	Emotion: sadness.	Strong emotion and affect: anger, crying.
Geographical scales	Intermediaries from local to global.	Speaking locally *and* globally.

Edited Media Witnessing: Mainstream Media Accounts of Indigenous Victim-Heroes and an Emerging Focus on Political Representation

During COP15, indigenous peoples appeared as victims of climate change when they were featured in a small number of Canadian and Swedish news articles (3 in Sweden, 4 in Canada) (Roosvall & Tegelberg, 2013). When indigenous peoples witness about climate change in these articles, a *victim-witness* role (Peters, 2001, p. 714) emerges. However, indigenous peoples are not only victims but also heroes since they are alerting other people to the urgency of climate change (Roosvall & Tegelberg, 2013). Thus they are not only victim-witnesses, but also victim-heroes that ultimately take on a victim-hero-witness role by alerting the rest of the world to the fact that something must be done. Hence, indigenous peoples function as intermediaries between nature and the West, mediating urgency, spirituality and de-nationalization (Roosvall & Tegelberg, 2013). The latter corresponds to how citizens increasingly voice concerns today: they do so not by rallying people around the nation, but rather by offering "… multiple perspectives on the conflict, calling into being a trans-national, rather than national, imagined community" (Chouliaraki, 2015b, p. 106, cf. Kampf & Liebes, 2013). Take, for example, a *Toronto Star* story that featured the image of an Inuit demonstrator with a caption that read: "Inuit, who say their way of life is threatened by climate change, demonstrate in Copenhagen on Thursday" (Woods, 2009). Here the multimodality of text and visual expresses what it is like to experience the effects of climate change—which we term *witnessing as experience*—while simultaneously acting as a catalyst for action—which we term *witnessing as speech act*.

In Sweden, the popular newspaper *Aftonbladet* ran a reportage series entitled "The World's Chance", focusing on affected peoples from all continents. Three articles make reference to indigenous peoples (see also Roosvall & Tegelberg, 2013). "Village collapsing into the sea" (Kerpner, 2009) covers the Inuit village Shishmaref, Alaska. The story features four Inuit who each witness climate change by recounting their experiences as victims of its devastating effects. The main witness is 62-year-old Ardith Weyiouanna. We learn that she grew up in a dugout house with a roof window on the tundra. In her testimonial she describes:

1) How safe the community was before climate change effects were felt: "You could not hear the wind, nor see the waves, and it was so safe";
2) How horrible it felt when effects threatened the safety of her home and eventually forced her to move: "… sometimes it was frightening. Water from storm waves poured down both sides of the house";
3) The fatal effects of climate change in the village: "It was so shocking. The ice was bad although it should not have been. It is the first casualty in our lifetime to take place in that way";
4) The impact climate change has on their ability to obtain food: "It is because the ice holds a shorter period of time, and the animals follow the ice." (Kerpner, 2009)

Stanley Tocktoo, a hunter and former mayor, witnesses about how female walruses are starting to abandon their cubs, adding that this too could be connected to the fact that there is less ice and more open water. Ken Stenek witnesses about spiritual experiences and climate change effects. He describes how his son Norman assumes the characteristics of another Norman who fell through the ice and died at a time when the ice normally should have held. This is what Ardith Weyiouanna alludes to when speaking about fatal effects. Stenek also testifies on his experiences with storms, waves and air pressure in and around his home. Ardith's husband Johnny Weyiouanna supports her testimonies about food scarcity and harsh living conditions, suggesting that if they cannot hunt animals like they used to the option of buying meat is out of the question because it is too expensive.

Lucia Quispe of the Aymara people in Bolivia is the main witness in the second article from "The World's Chance" series (Jönsson, 2009). She says that "When he [her 3-year-old son] is grown up, this village will probably not exist anymore." Like the people in Shishmaref, Alaska, Quispe and other people from Mount Illimani, Bolivia witness about food shortages caused by climate change. In another quote, Quispe refers to the crops she is about to sell at the market, stating that "Six months work and this is all. It will not be enough to make our yearly purchases in town." The story subsequently connects her witnessing of specific changes to the snow cover on the mountain to statements expressing a spiritual view of nature: "We talk to the mountain to check how it fares. Now the mountain is dying …" The author later explains that when the weather began changing and the usual rains did not come, the people's first impression was that they had made the wrong offerings to Pachamama (Mother Earth): "They thought that the Gods took revenge by not granting them water. It was only when a water engineer came to the village that the villagers learned that the region was affected by something called global warming" (Jönsson, 2009). Indigenous knowledge is thus portrayed as likable, but ultimately wrong; evidenced by how local beliefs are corrected by a water engineer. The article includes pictures of cracked soil where there was once a water reserve, and of Quispe and others wearing traditional hats that are mentioned in the text.

The final article about indigenous peoples in the *Aftonbladet* series on climate change impacts in local settings around the world covers the Sami (Lindberg, 2009). Jacob Nygård, a reindeer herder, is the main witness. He testifies on what happened five weeks prior when 291 reindeer fell through the ice and died, at a time when the ice on the lake was supposed to hold. Nygård is quoted saying: "I thought, they have to make it. Every one of them have to get to the shore safely." The article describes how Nygård used to travel with his grandfather up the mountain where he learned to understand the reindeer, how they think and move. The author explains that "Lately, the disaster has been in focus a lot. Not concerning the money that was lost, but concerning the animals, that he has lived with for so long". Nygård then says: "I am sorry they had to suffer … It hurts." The article includes pictures of Jakob Nygård and the dead reindeer. While the aforementioned articles reported that help with infrastructure was needed, this one represents specific Sami opinions on *what* infrastructure should be developed and *where*. The Sami leet (parliament) is mentioned in a small fact box.

In the 2009 articles from Swedish mainstream media, the combination of witnessing as experience and witnessing as speech act constitutes strong messaging. Indigenous voices are not only included, but centred, giving them a key role in the stories. Contextual factors can downplay this, for example by deeming certain testimonies false or inferior compared to statements from scientists, as in the article about the Aymara people. The testimonies are all emotional, expressing sadness and explaining how it feels. The indigenous people who speak are witnessing changes in weather and nature—the effects of climate change in their localities. These effects are dramatic to varying degrees, including the lack of access to food and the death of humans and animals. Furthermore, these witnesses express concern about their villages and fears that their sustainable ways of life will be extinguished. The issues they testify about thus concern their very existence, and thereby connect closely to human rights and the right to have rights, even though this is not brought up explicitly in the articles. The urgency of climate change is mediated from indigenous peoples who appear as intermediaries to the rest of the world. However, the urgency of action that would help save their own communities is concealed by the fact that political perspectives are not the focus. This seems to be connected to how spiritual and political perspectives tend to be mutually exclusive (Roosvall & Tegelberg, 2012, 2013).

Bjørst (2012) has noted similar tensions in Inuit testimonies within the framing of global climate crisis narratives, where witnessing on the global and the local do not align. She observes how being a global witness enhances the victim role while being a local witness emphasizes agency. In this sense, the indigenous witnesses appear to be speaking in the name of a we when speaking to save the planet (or speaking on behalf of the climate, as Bjørst concludes) and as a we when trying to make connections between the global and the acute situations in their own local communities, directing political action back to them. It is noteworthy that some versions of the examined articles were accompanied by fact boxes where affected people could pose questions to politicians at COP15. These constructive and creative fact boxes made connections to the political realm. They were nonetheless separated from the main articles, only appearing in some versions, without allowing indigenous peoples to provide answers. They could merely pose questions to others who provided answers. Political *representation* is not advocated in the series and there is no political deliberation in the articles; the level of participation is mainly dialogue.

The Swedish mainstream media material from 2015 differs from the 2009 material since political representation is discussed explicitly. It is addressed in three articles published in the printed version of *Dagens Nyheter*, and in one published exclusively online. We zoom in on one example, which includes several expressions of witnessing. In "The Sami excluded from climate negotations in Paris" (Larsson, 2015), Jannie Staffansson and Per Jonas Partapuoli (who are also both interviewees in this study), appear as witnesses together with the Vice Chair of the Swedish Sami leet, Josefina Lundgren Skerk. They are each quoted as follows:

> We are melting. The only reason I can be here is because we have not been able to start rounding-up our reindeer and move them down to winter land. Ices have not frozen so we cannot herd them, tells 22-year-old reindeer herder Per Jonas Partapuoli, who lives outside of Kiruna.

> It [the fact that no Sami representative is included in the Swedish negotiating delegation] affects our right to self-governance. It is very negative because our future is at stake, says 25-year-old Jannie Staffansson, who tries to influence Swedish politicians on location.

> It is scandalous that we aren't included like that [like the Norwegian and Finnish Sami]. The Sami leet is both a parliament elected by popular vote and a public authority. It should be a given that we are represented, says Josefina Lundgren Skerk, vice chair of the Sami leet, representing the Sami Hunting and Fishing Party.

Like the 2009 articles, this one contains witnessing about weather and nature, but unlike them it is combined with witnessing on politics and injustice, albeit rather briefly. Like the 2009 articles, spiritual and political aspects are not combined; here, however, spiritual aspects are eschewed while political ones are emphasized. Emotion is more implicit than in the 2009 material, appearing through word choices like "scandalous" and "negative."[15] The *joint* tendencies in the edited media witnessing can be clarified further by detailing examples of un-/self-edited media witnessing.

Un-/Self-Edited Media Witnessing: The Political, Pluralism and the Bridging of Discourses

In a video posted on YouTube, Jannie Staffansson speaks at a COP21 event (December 11) with, among others, *Greenpeace's* Kumi Naidoo.

> My friends, this is the face of a *angry* Sami person. Indigenous peoples, that are not represented under the negotiations, we are being *silenced*, by those states who speak up for us, those states who put forward language on our *rights*, *human rights*, *indigenous peoples rights*. Multiple times going to the Secretariat of France, complaining by those changes not being edited within the draft of negotiations agreement, and then being *silenced* again. We are very critical to the actions of France, how they continuously *silence* the persons who are not even here. *We are the persons who are dying*. My friends,

> my family, they are the ones who go through water, they are the ones who are *killed in avalanches* [holding back tears, starts crying]. And they don't even care. *How can the purpose of this negotiation not be people?* How can our *voices* be *silenced*, multiple times, and then again? I'm sorry my friends, but we need *civil society* to rise up. We have done what we can under these negotiations. And we now call for civil society to rise. Push your ministers! We need human rights, indigenous peoples' rights in that *agreement* [still emotional] in order to stop violations, killings, murders, land grabbing, and to save the planet for our future. Thank you. (Månsson, 2015)

Drawing out significant themes and concepts (in italics) and analysing how they are articulated and related to each other,[16] makes it clear that this un-/self-edited media witnessing places more emphasis on the political, and explicitly talks about (lack of) voice and silencing as part of a political process. Staffansson also connects to the need for a focus on the who of justice, asking: "How can the purpose of this negotiation not be the people?" She stresses that she is angry, and eventually starts crying, thereby highlighting emotion (speaking about feelings—"angry") and displaying affect (crying) (see Papacharissi, 2014; Shouse, 2005; Wahl-Jørgensen, 2016).

In a video of the prayer ceremony, detailed in the introduction to this book, Eagle Woman and Dallas Gooldtooth, both Native American activists affiliated with the Indigenous Environmental Network, appear as witnesses. The video was displayed on the Indigenous Environmental Network's Facebook page and later picked up by *CBC News—Indigenous* in Canada. The video mixes chanting from the prayer ceremony with testimonies delivered directly into camera or as voice-over. In the following passage each of the witness statements are quoted, with significant word choices and displays of emotion or affect italicized.

Eagle Woman:	[Voice-over] Every time we go into a new space as indigenous peoples we open with a *blessing* [cut to Eagle Woman close-up interview] and we want to stand in *solidarity* [*crying*] with people and let them know that we're here and we care. And to send a message of love [cut to panning the flowers surrounding the park, as remembrance of those who died at the Bataclan across the street] and hope and peace for the future.
Dallas Goldtooth:	[Voice-over] We as impacted *frontline communities* are quite familiar with *tragedy*, we understand [cut to Dallas Goldtooth chanting in the ceremony] what it means to have *great loss*.
Eagle Woman:	Our people are now dealing with, in *my community*, with fracking, and coal, at the same time, we have coal fire power plants that have *made our* [*crying*] *babies sick*.

Dallas Goldtooth: The *system* as a whole needs to change, in order for us to really see any lasting peace on this *planet*.

Eagle Woman: We're *fighting* for a *systems change* [*crying*], and it will happen, and we're not going to ever quit until it does. It's just a matter of when. (Indigenous Environmental Network, 2015)

These statements are followed by the text: "Paris Climate Talks. Reporting from the frontlines of Climate *Justice*" (emphasis added). Each testimony combines spiritual (blessing, chanting) with radical political (systems change, fighting) critique, in a manner that remains lacking in mainstream media coverage of indigenous peoples and climate change (see Roosvall & Tegelberg, 2013). The witnessing also includes a bridging of local experiences in their own communities with the global (planet). Experiences of tragedy, loss and sick babies are expressed with emotion coming across stronger (e.g. crying) than it does in mainstream media accounts.

In the un-/self-edited format, whether indigenous witnesses are representing ordinary people or activists, they speak from a non-professional perspective in terms of journalism. Hence, in this sense they are performing "ordinary witnessing" rather than "journalistic witnessing" (Chouliaraki, 2011). The Indigenous Environmental Network is however ambiguous in this regard, since it features people who are activists that are working to produce their own media. This epitomizes the rising phenomenon of activists as news producers or journalism as activism (see Breindl, 2016; Russell, 2016). In the following we elaborate on conclusions that can be drawn from the analysis of diverse forms of media witnessing and discuss these in relation to more general conclusions, drawing on the whole scope of the chapter.

Conclusion

Edited media witnessing is more geared towards consensus, whereas un-/self-edited witnessing is more agonistic, stressing conflict and the need for systems change. The latter includes testimony that evokes issues of misrecognition, maldistribution and misrepresentation (political) (Fraser, 2008). The fact that edited witnessing does not include the same complexity of references to injustice may be due to the fact that edited witnessing is constituted as reactions—both to reporters' questions and to climate change as such. By contrast, un-/self-edited witnessing is more action-oriented. It includes testimonies on the level of political deliberation, not just dialogue—which dominates the edited media witnessing material from 2009, or mere presence which characterizes the other media material that features indigenous peoples. Political agency is thus underlined more in un- and self-edited media material. Sometimes this occurs in conjunction with cultural and spiritual agency. In edited media witnessing, the agency of indigenous peoples is *conditional* (Chouliaraki, 2006, pp. 119, 124). It occurs as reactions to questions posed by journalists and appears within certain frames, which often downplay political agency while stressing cultural and spiritual agency. In other words, indigenous peoples tend to appear conditionally as agents when they speak of spirituality or of climate change in general ("non-politically").

Witnessing in legacy media regularly constitutes publics by speaking "objectively" *in the name of* a "we". "Ordinary witnessing," outside legacy media, often constitutes publics by *positioning themselves as* the "we" (Chouliaraki, 2011, p. 308), with people "being represented by themselves" (Atton, 2002, p. 122, cf. Chouliaraki, 2011). This difference is epitomized by indigenous peoples witnessing during the COPs. In edited witnessing, indigenous peoples become intermediaries, messengers speaking on behalf of a we. In un-/self-edited material, they speak for themselves *as* a we, focusing more on what must be done for indigenous groups instead of simply appearing as victim-witness-heroes who inform others of the global urgency of climate change and about how to view nature. The un-/self-edited material tends to *merge* geographical scales, simultaneously speaking locally *and* globally (see also Bjørst, 2012; Stephansen, 2013) unlike edited media witnessing stories where mediating *between* the local and the global (see also Stephansen, 2013, p. 103) is commonplace. Thereby the un-/self-edited material constructs "alternative knowledge projects" (Stephansen, 2013), connecting to an imagining of alternative worlds which defines an agonistic cosmopolitics of human rights (Hanafin, 2012). In addition, the un-/self-edited material is generally more pluralistic. Hence, it relates more closely, once again, to agonistic democracy and an agonistic cosmopolitics of human rights. In some instances, this pluralism includes connections between spirituality and politics, which are uncommon in the edited media witnessing material.

Another difference between the edited and un-/self-edited witnessing is a stronger emphasis on emotion and affect in the latter. This could arguably be explained by the fact the material was presented on video rather than in texts and still photographs. However, texts and still images are capable of describing emotions as well. In any case, the combination of what indigenous peoples witness *about* and *how* they go about it emphasizes the application of "political emotions" (Nussbaum, 2013) in the un-/self-edited material.

Overall we found striking parallels between representation in politics and in the media. Indigenous activists struggle in both these arenas to heighten their level of access in order to see and be seen in the public realm as an agonistic space of appearance. This occurs through "lobbying", as Lene Kielsen Holm puts it, the relevant states (in her case Greenland and Denmark); public speaking—which occasionally finds its way to YouTube (as in the witnessing of Jannie Staffansson); through indigenous produced or co-produced media, as in the case of the Indigenous Environmental Network (IEN) and Amazonwatch; and through edited media witnessing and public speaking in the mainstream media. Although the "bigger story" is still not, as Allison Akootchook Warden puts it, about collective actions and relationship building outside the COPs, the mainstream media does occasionally connect the harshly felt consequences of climate change in indigenous communities to the politics of the meetings, as exemplified by the 2009 stories in Swedish and Canadian media. In 2015, a number of Swedish stories were moreover explicitly focused on voices from the "outside" trying to get "inside." These stories deal with political participation or rather the lack of it, which thus invites a discussion about misrepresentation (Fraser, 2008), and the very right to have rights. To be able, in other words, to actually realize and practice human rights (Arendt, 1998; Oman, 2010). Furthermore it is noteworthy that in 2015 Sami and other indigenous groups *did* have a voice in one "elite" Swedish newspaper, *Dagens Nyheter*.[17]

While discourses on injustice concerning political representation are starting to find their way into the mainstream media, they are expressed in much stronger and more elaborated form in the un-/self-edited media witnessing. There we find expressions of political emotion (see also Andén-Papadopoulos, 2014, p. 761), which are not hugely surprising given the known tension between the historical paradigm of objectivity in professional journalism and the more emotionally charged manner of citizen journalism (Allan, 2013; Wahl-Jørgensen, 2016, pp. 13–14). However, these differences are still noteworthy, firstly due to the importance of affect in agonistic democracy. And, secondly, because the more affective dimensions of un-/self-edited media witnessing texts connect to justice by evoking political representation. The challenge for legacy news media may thus be, as Beckett and Deuze (2016, p. 1) note, to "become a meaningful, insightful and trustworthy part of an emerging affective media ecosystem."

A crucial question is ultimately who is listening to witnessing, public speaking and other modes of appearing in the public realm. Who "listens" to a testimony on YouTube? Sometimes a significant number of people, sometimes not. The video discussed here featuring Jannie Staffansson had 817 views in March 2017.[18] The video produced by the IEN, in turn, includes the statement "The mainstream media ignored this but yesterday indigenous groups held a healing ceremony" (CBC News, 2015). Apart from being displayed on the IEN Facebook page, the video appeared on the website for *CBC News* "Indigenous" in Canada. In March 2017, it had been viewed over 1 million times on Facebook. Regardless of the attention generated on Facebook, there was a wish among its producers that the mainstream media would pay attention to this and other indigenous climate issues. It is thus not only a question of *how many people* listen but *who* these people are. From this perspective, Jannie Staffansson's brief appearance in an elite Swedish newspaper may be crucial in terms of reaching politicians and elites within the nation-state where her indigenous community dwells. However, as Russell (2016, p. 8) points out, there is indeed a continuum between mainstream and alternative media today, and we argue that they both still have crucial roles to play, albeit in different ways.

Regardless of media platform, in a mediatized world it is necessary to see and hear the arguments of an adversary in order for agonistic democracy to work. The connected need to formulate your own arguments depends largely on media articulation in conjunction with the absolute necessity of recognition and representation in politics. It is not merely the fact one can formulate arguments, but *how* one formulates them that matters. Which modes of communication are possible in what arenas? Can you express political emotions? Can political, cultural and spiritual discourses co-exist? Can you speak—and are you spoken to—in a mode of solidarity, in terms of a collective responsibility? We will elaborate on the latter question in the next chapter, where we also discuss how the creation of a Common Human Identity (see Chapter 2) might be possible in relation to the particularistic demands posed by indigenous peoples. It is, as we shall see, a question of recognizing identity versus status.

Notes

1. For a more detailed discussion of our methodological approaches, see Chapter 1; when needed, specifics are clarified here.
2. For a detailed account of the status granted to different types of actors at UN climate summits, see Chapter 3.

3. For further elaboration on access (presence), dialogue and deliberation, see Roosvall and Tegelberg (2012).
4. The Swedish newspaper *Aftonbladet* included the only article about a non-Arctic indigenous group, detailed in the section on media witnessing.
5. In the section on media witnessing we elaborate on these levels of *dialogue* and *political deliberation*, relating them to indirect vs. direct participation. There we distinguish between mainstream media, where indigenous actors are framed by journalists, and social, self-produced media where they speak more directly to the public. In this section we also consider articles from 2015 where the level of political deliberation increases.
6. See Chapter 2 for a fuller elaboration of the politics of reframing.
7. This differs from Mouffe's (2013) understanding of identity and difference in the agonistic order.
8. For more on journalistic framings of politics as a strategic game, rather than one of substance or issues, see Strömbäck (2005).
9. The frame analysis focused on headlines, but also included images, captions and other conspicuous aspects of the articles, in accordance with Entman's (1993, p. 52) emphasis on the importance of selection and salience in framing processes.
10. These two main frames also interacted in 27 articles (Roosvall, 2010, p. 314).
11. The voice data in this figure was collected over three weeks before, during and after COP15 in 2009 (Copenhagen), COP17 in 2011 (Durban), COP19 in 2013 (Warsaw) and COP21 in 2015 (Paris). The sample includes 7287 articles from two newspapers in 11 countries (Australia, Bangladesh, Canada, Egypt, Finland, Israel, Norway, Russia, South Africa, Sweden, USA). We are grateful to the MediaClimate research network for sharing this data with us, and to Elisabeth Eide, Risto Kunelius and Dmitry Yagodin especially for providing us with this figure. For more background on the MediaClimate network, see https://mediaclimate.net/. Information on the newspapers collected from each country can be found in Eide & Kunelius (2012).
12. See also Pedelty, 1993 (cf. Bishara, 2012, p. 143) on how indigenous culture is essentialized and political messages of solidarity watered down, partly by focusing on traditional clothing in US photojournalism of indigenous groups in Guatemala.
13. For further discussion of this movement, see Chapter 2.
14. Here is the link that Näkkäläjärvi provided to the Yle Sápmi broadcast featuring the news anchor's yoik: http://yle.fi/uutiset/osasto/sapmi/gulahallat_eatnamiin_-luohti_cuodja_yle_oasiin/8421339.
15. Some of these differences may be attributed to the fact that the material from 2009 is from a popular newspaper whereas the 2015 material is from an elite newspaper. It is also noteworthy that the Sami had a voice in the elite Swedish newspaper in 2015, while only having a voice in the popular newspaper earlier.
16. This is done using the method of discourse theory, focusing on articulations of elements of discourse and chains of equivalence (Laclau & Mouffe, 1985; Winter-Jørgensen & Phillips, 2002).
17. In the previous years we examined (2009, 2011, 2013), Sami and other indigenous peoples were only featured on the level of presence in the Swedish elite newspaper (2011). They only had a voice in the popular newspaper, *Aftonbladet* in 2009 and 2011.
18. An audio excerpt of the same speech had been played 803 times in December 2016.

References

Allan, S. (2013). *Citizen witnessing: Revisioning journalism in times of crisis*. London: Polity.

Amazon Watch. (2015a, December 3). Humanity—together—is in danger of extinction. Are we really understanding this? *Amazonwatch.org*. Retrieved from http://amazonwatch.org/news/2015/1203-humanity-together-is-in-danger-of-extinction-are-we-really-understanding-this

Amazon Watch. (2015b, December 8). Amazonian "Canoe of Life" brings indigenous rights message to COP21: Indigenous hand-carved canoe from Ecuadorian Amazon arrives in Paris. [Press release] *Amazonwatch.org*. Retrieved from http://amazonwatch.org/news/2015/1208-amazonian-canoe-of-life-brings-indigenous-rights-message-to-cop21

Amazonwatch.org. (2015c, December 5). Indigenous flotilla of kayaks and press conference demanding true climate solutions at COP21 [Press release]. *Amazonwatch.org*. Retrieved from: http://amazonwatch.org/news/2015/1205-indigenous-flotilla-of-kayaks-and-press-conference-demanding-true-climate-solutions-at-cop21

Andén-Papadopoulos, K. (2014). Citizen camera-witnessing: Embodied political dissent in the age of "mediated mass self-communication". *New Media & Society*, *16*(5), 753–769.

Arendt, H. (1951). *Origins of totalitarianism*. New York, NY: Harcourt, Brace, Janovitch.

Arendt, H. (1998). *The human condition* (2nd ed.). Chicago, IL: University of Chicago Press.

Askanius, T., & Uldham, J. (2011). Online social media for radical politics: Climate change activism on YouTube. *International Journal of Electronic Governance*, 4(1), 69–84.

Atton, C. (2002). *Alternative media*. London: Sage Publications.

Bagelman, J. (2016). Geo-politics of paddling: 'Turning the Tide' on extraction. *Citizenship Studies*, *20*(8), 1012–1037.

Balibar, E. (2002). *Politics and the other scene*. London: Verso.

Beckett, C., & Deuze, M. (2016). On the role of emotion in the future of journalism. *Social Media + Society*, *2*(3), 1–6.

Birmingham, P. (2007). The an-archic event of natility and "the right to have rights. *Social Research*, 74(3), 763–776.

Bishara, A. (2012). Circulating the stances of liberation politics: The photojournalism of the anti-wall protests. In M. McLagan, & Y. McKee (Eds.), *Sensible politics: The visual culture of nongovernmental activism* (pp. 139–148). New York, NY: Zone Books.

Bjørst, L. R. (2012). Climate testimonies and climate crisis narratives: Inuit delegated to speak on behalf of the climate. *Acta Borealia*, *29*(1), 98–113.

Breindl, Y. (2016). Activists as news producers. In T. Witschge, C. W. Anderson, D. Domingo, & A. Hermida (Eds.), *The SAGE handbook of digital journalism*. London: Sage Publications.

Cable, J. (2016). *Protest campaigns, media and political opportunities*. London: Rowman & Littlefield International.

Cammaerts, B., Mattoni, A., & McCurdy, P. (2013). Introduction: Mediation and protest movements. In B. Cammaerts, A. Mattoni, & P. McCurdy (Eds.), *Mediation and protest movements* (pp. 1–20). Bristol: Intellect.

Cappella, J. N., & Jamieson, K. H. (1997). *Spiral of cynicism: The press and the public good*. New York, NY: Oxford University Press.

Cavarero, A. (2000). *Relating narratives: Storytelling and selfhood*. London: Routledge.

Cavarero, A. (2002). *Stately bodies: Literature, philosophy, and the question of gender*. Ann Arbor, MI: University of Michigan Press.

Cavarero, A. (2005). *For more than voice: Toward a philosophy of vocal expression*. Stanford, CA: Stanford University Press.

CBC News. (2015, November 30). Indigenous leaders, activists in Paris for climate change talks. *CBC Radio Canada*. Retrieved from http://www.cbc.ca/news/indigenous/indigenous-leaders-activist-paris-climate-talks-1.3343927

Chambers, S. A. (2001, December 11). Language and politics. Agonistic discourse in the west wing. *CTheory*. Retrieved from http://www.ctheory.net/articles.aspx?id=317

Chandler, D. (2009). Critiquing liberal cosmopolitanism? The limits of the biopolitical approach. *International Political Sociology*, 3(1), 53–70.

Christensen, M., Nilsson, A. E., & Wormbs, N. (Eds.). (2013). *Media and the politics of arctic climate change: When the ice breaks*. Basingstoke: Palgrave.

Chouliaraki, L. (2006). *The spectatorship of suffering*. London: Sage.

Chouliaraki, L. (2011). Improper distance: Towards a critical account of solidarity as irony. *International Journal of Cultural Studies*, *14*(4), 363–381.

Chouliaraki, L. (2013). *The ironic spectator: Solidarity in the age of post-humanitarianism*. Cambridge, MA: Polity.

Chouliaraki, L. (2015a). Digital witnessing in war journalism: The case of post-Arab Spring conflicts. *Popular Communication*, *13*(2), 105–119.

Chouliaraki, L. (2015b). Digital witnessing in conflict zones: The politics of remediation. *Information, Communication & Society*, *18*(11), 1362–1377.

Christie, M. (2001). From civil to political religion: Intersections of culture religion and politics. Waterloo, ON: Wilfrid Laurier University Press.

Clark, B. (2002). The indigenous environmental movement in the United States: Transcending borders in struggles against mining, manufacturing, and the capitalist state. *Organization & Environment*, *15*(4), 410–442.

Cocq, C. (2016). Exploitation or preservation? Your choice! Digital modes of expressing perceptions of nature and land. In H. Graf (Ed.), *The environment in the age of the internet: Activists, communication and the digital landscape* (pp. 53–72). Cambridge: Open Book Publishers.

Connolly, W. E. (2005). *Pluralism*. Durham, NC: Duke University Press.

Couldry, N. (2009). Commentary: Rethinking the politics of voice. *Continuum. Journal of Media and Cultural Studies*, 29(4), 579–582.

Curran, J. (2011). *Media and democracy*. Abingdon: Routledge.

Eide, E., & Kunelius, R. (2012). Introduction. In E. Eide, & R. Kunelius (Eds.), *Media meets climate: The global challenge for journalism* (pp. 297–312). Gothenburg: Nordicom.

Elster, J. (1998). Introduction. In J. Elster (Ed.), *Deliberative democracy* (pp. 1–18). New York, NY: Cambridge University Press.

Entman, R. (1993). Framing: Toward clarification of a fractured paradigm. *Journal of Communication*, *43*, 51–58.

Fenton, N. (2016). *Digital, political, radical*. Cambridge: Polity.

Fishkin, J. (1995). *The voice of the people: Public opinion and democracy*. New Haven, CT: Yale University Press.

Fraser, N. (2008). *Scales of justice: Reimagining political space in a globalizing world*. New York, NY: Columbia University Press.

Frosh, P., & Pinchevski, A. (Eds.). (2009). *Media witnessing: Testimony in the age of mass communication*. London: Palgrave Macmillan.

Gutman, A., & Thompson, D. (2009). *Why deliberative democracy?* Princeton, NJ: Princeton University Press.

Habermas, J. (1994). Three normative models of democracy. *Constellations*, *1*(1), 1–10.

Hanafin, P. (2012) A right to politics? Towards an agonistic cosmopolitics of human rights. In G. Delanty (Ed.), *Routledge handbook of cosmopolitanism studies* (pp. 326–339). Abingdon: Routledge.

Honig, B. (2009). *Emergency politics: Paradox, law, democracy*. Princeton, NJ: Princeton University Press.

Husband, C. (2000). Media and the public sphere in multi-ethnic societies. In S. Cottle (Ed.), *Ethnic minorities and the media: Changing cultural boundaries* (pp. 199–213). Buckingham: Open University Press.

Indigenous Environmental Network. (2015, November 30). *Facebook* [Group page]. Retrieved from https://www.facebook.com/ienearth/videos/10153285239180642/

Jönsson, H. B. (2009, December 18). Our mountain is dying [Vårt berg är på väg att dö]. *Aftonbladet*.

Kampf, Z., & Liebes, T. (2013). *Transforming media coverage of violent conflicts: The new face of war*. London: Palgrave Macmillan.

Kaun, A. (2016). *Crisis and critique: A brief history of media participation in times of crisis*. New York: Zed books.

Keck, M. E., & Sikkink, K. (1998). *Activists beyond borders: Advocacy networks in international politics*. Ithaca, NY: Cornell University Press.

Kerpner J. (2009, December 16). Village collapsing into the sea [Byn rasar i havet]. *Aftonbladet*.

Laclau, E., & Mouffe, C. (1985). *Hegemony and socialist strategy: Towards a radical democratic politics*. New York, NY: Verso.

LaDuke, W. (2005). *Recovering the sacred: The power of naming and claiming*. Cambridge, MA: South End.

Lightfoot, S. (2016). *Global indigenous politics: A subtle revolution*. Abingdon: Routledge.

Lindberg, S. (2009, December 19). The ice does not hold anymore: How climate change is affecting Sweden [Isen bär inte längre. Så slår klimatförändringarna—i Sverige]. *Aftonbladet*.

Macnamara, J. (2016). *Organizational listening: The missing essential in public communication*. New York, NY: Peter Lang.

Månsson, H. (2015, December 11). Jannie Staffansson and Kumi Naidoo at COP21, Paris. *YouTube*. Retrieved from https://www.youtube.com/watch?v=BnT7l-haRr8&t=8s

Mouffe, C. (2005). *On the political*. London: Routledge.

Mouffe, C. (2013). *Agonistics: Thinking the world politically*. London: Verso.

Nash, K. (2014). Introduction. In K. Nash (Ed.), *Transnationalizing the public sphere* (pp. 1–7). Cambridge: Polity Press.

Nussbaum, M. (2013). *Political emotions: Why love matters for justice*. Cambridge, MA: Belknap, Harvard University Press.

Larsson, M. (2015, December 3). Samerna utestängda från klimatförhandlingarna i Paris. *Dagens Nyheter*. Retrieved from http://www.dn.se/nyheter/varlden/samerna-utestangda-fran-klimatforhandlingarna-i-paris/

Oman, N. (2010). Hannah Arendt's "Right to have rights": A philosophical context for human security. *Journal of Human Rights*, 9, 279–302.

Painter, J. (2010). *Summoned by science: Reporting climate change at Copenhagen and beyond*. Oxford: Reuters Institute for the Study of Journalism.

Papacharissi, Z. (2014). *Affective publics: Sentiment, technology and politics*. New York, NY: Oxford University Press.

Peters, J. D. (2001). Witnessing. *Media, Culture and Society*, *23*(6), 707–723.

Pietikäinen, S. (2003). Indigenous identity in print: Representations of the Sami in news discourse. *Discourse and Society*, *14*(5), 581–609.

Plaut, S. (2012). "Cooperation is the story"—Best practices of transnational indigenous activism in the North. *The International Journal of Human Rights*, 16(1), 193–2015.

Roberts-Miller, P. (2002). Fighting without hatred: Hanna Arendt's agonistic rhetoric. *JAC: A Journal of Composition Theory*, *22*(3), 585–601.

Roosvall, A. (2010). Sweden: Between domestication and glocalisation. In E. Eide, K. Kunelius, & V. Kumpu (Eds.), *Global climate, local journalisms: A transnational study of how media make sense of climate summits* (pp. 309–324). Bochum: Projekt Verlag.

Roosvall, A. (2011). What is threatening the West? Islam/communism, religion/politics and the rational/irrational discourse. In S. Nohrstedt (Ed.), *Communicating risk: Towards the threat society?* (pp. 46–97). Göteborg: Nordicom.

Roosvall, A. (2016). Journalistik, rättvisa och rättigheter. *Människorna, medierna och marknaden: Medieutredningens forskarantologi om en demokrati i förändring*. Stockholm: Statens Offentliga Utredningar.

Roosvall, A., & Tegelberg, M. (2012). Misframing the messenger: Scales of justice, traditional ecological knowledge and media coverage of indigenous peoples and climate change. In E. Eide & R. Kunelius (Eds.), *Media meets climate: The global challenge for journalism* (pp. 297–312). Gothenburg: Nordicom.

Roosvall, A., & Tegelberg, M. (2013). Framing climate change and indigenous peoples: Intermediaries of urgency, spirituality and de-nationalization. *International Communication Gazette*, *75*(4), 392–409.

Roosvall, A., & Tegelberg, M. (2015). Media and the geographies of climate justice: Indigenous peoples, nature and the geopolitics of climate change. *TripleC: Communication, Capitalism & Critique*, *13*(1), 39–54.

Russell, A. (2016). *Journalism as activism: Recoding media power*. Cambridge: Polity.

Sápmi. (n.d.). *Jojken är minnets tradition*. Retrieved from http://www.samer.se/1201

Segerberg, A., & Bennett, L. (2011). Social media and the organization of collective action: Using Twitter to explore the ecologies of two climate change protests. *Communication Review, 14*, 197–215.

Shouse, E. (2005). Feeling, emotion, affect. M/C *Journal*, 8(6). Retrieved from http://journal.media-culture.org.au/0512/03-shouse.php

Stephansen, H. C. (2013). Starting from the Amazon: Communication, knowledge and politics of place in the World Social Forum. *Interface: A Journal for and about Social Movements*, 5(1), 102–127.

Strömbäck, J. (2004). *Den medialiserade demokratin. Om journalistikens ideal, verklighet och makt.* Stockholm: SNS Förlag.

Strömbäck, J. (2005). In search of a standard: Four models of democracy and their normative implications for journalism. *Journalism Studies*, 6(3), 331–345.

Tegelberg, M. (2010). Canada: The dirty old man of climate politics? In E. Eide, & R. Kunelius (Eds.), *Global climate, local journalisms: A transnational study of how media make sense of climate summits* (pp. 97–114). Freiburg: Projek Verlag.

Tuchman, G. (1978). Introduction: The symbolic annihilation of women by the mass media. In G. Tuchman, A. K. Daniels, & J. Benet (Eds.), *Hearth and home: Images of women in the mass media* (pp. 3–38). New York, NY: Oxford University Press.

Wahl-Jørgensen, K. (2016). Emotion and journalism. In T. Witschge, C. W. Anderson, D. Domingo, & A. Hermida (Eds.), *The SAGE handbook of digital journalism* (pp. 128–144). London: Sage Publications.

Winter-Jørgensen, M., & Phillips, L. (2002). *Discourse analysis as theory and method*. London: Sage.

Woods, A. (2009, December 11). Making climate change personal. *Toronto Star*. Retrieved from https://www.thestar.com/news/world/2009/12/11/making_climate_change_personal.html

Young, I. M. (2013). *Responsibility for justice*. Oxford: Oxford University Press.

· 6 ·

(DIS)CONNECTIONS

Particularism Versus Universalism, and Transnational Solidarity

> My long-term goal is … to bridge indigenous groups because of the collective colonial experience … there are some local grassroots movements that are doing incredible stuff, really engaging in the political process in a way that still means an integrity with their own personal views … I think because of the collective experiences and struggles that there's a great opportunity to learn from that and bring it back to … grassroots community-based action.
>
> —India Logan-Riley, Māori, New Zealand (COP21)

India Logan-Riley addresses *connections* she wants to make in the context of colonial experiences and the struggles performed by colonized groups. *Disconnections* are also presupposed: disconnections between indigenous groups and full rights to their native lands, as well as disconnections of their voices from political processes (see Chapter 5). Many interviewees, Logan-Riley among them, also expressed concerns regarding disconnections between indigenous voices and dominant national and international media (Chapter 4). Logan-Riley simultaneously relates to a bridging of *the particular and the universal.* This is of specific interest in relation to indigenous groups given that they are the subjects of both particular indigenous and universal human rights. This bridging can help us understand how to tackle the particular and the universal of climate change, a problem that is global in scope but has graver consequences in specific localities; not least in those where indigenous peoples live (Chapter 3).

In addressing these (dis)connections, Logan-Riley evokes a discourse of *transnational solidarity* that concerns itself with issues of justice.

This book has examined the transnational dimensions of climate justice. Focusing on UN climate summits, we have illuminated some of the challenges faced by transnational indigenous activists who attend international climate summits to advocate on behalf of their local interests while simultaneously making broader demands for climate action. Hence, we link the case of transnational indigenous peoples to broader debates concerning climate justice, geographical scales, media ecologies, activism and politics. The book revealed significant (dis)connections in each of these areas. Regarding understandings of *climate justice* (Chapter 2), it unveiled disconnections between diverse geographical scales; exposing an almost exclusive focus on the international scale, which excludes transnational peoples and disregards repercussions inherent in the transnational character of climate change as such. It also showed how general ideas concerning responsibility for justice must be connected to how the media works, based on the hardships indigenous peoples meet when they struggle to have their voices heard in the media.

Regarding the *geopolitics* of local climate change effects and of the global climate summits as political spaces (Chapter 3), the book exposed disconnections between political "centres" and "peripheries", which manifested themselves at the summit in the (limited) structures of access for indigenous peoples. Concurrently, indigenous activists from around the world worked together inside and outside that structure, thus connecting people transnationally and even globally.

Regarding the *media ecologies* in which summit reporting, media strategizing, media production and consumption take place (Chapter 4), we found an increasing disconnection between indigenous activists and the mainstream media. This was specifically manifested by a mistrust many interviewees expressed based on experiences of unjust and unknowledgeable reporting, which subsequently caused many to avoid consuming any mainstream media. We also found a disconnection between indigenous media and mainstream media. Had they instead been connected, this could have served to amend the lack of knowledge concerning indigenous issues that mainstream media often display. Additionally, we found disconnections between social media and mainstream media in terms of sourcing and voices represented in media output on indigenous peoples and climate change.

Regarding *activism* and *politics* (Chapter 5), we demonstrated how edited media witnessing, done by activists as well as indigenous peoples in mainstream media, disconnected political perspectives from indigenous peoples and climate change; while un-/self-edited media witnessing published by indigenous organizations or partners in social media connected discourses on politics, spirituality and culture. We also delved deeper into issues of access to political processes at the summits, showing that indigenous representatives are generally disconnected from decision-making, and that the process of inclusion/exclusion often appears as rather arbitrary, characterized, for instance, by sudden withdrawals of access. Indigenous representatives counter these processes by establishing and implementing connections with other indigenous groups who may happen to have more access in certain contexts.

Disconnections are not necessarily bad as such. Diverse media can serve diverse purposes, thereby quite naturally appearing at a distance from one another. However, in relation to climate change, one of the main global crises of our times (Cottle, 2009), more connections would certainly be constructive. Particularly, connections between media that help to facilitate: the inclusion of a plurality of voices and perspectives, including those of indigenous peoples; better access to powerful elites in the decision-making arena; and the capacity to reach wider audiences. We shall elaborate on these ideas further at the end of this chapter.

This chapter is divided into three parts corresponding to the three aforementioned themes: (dis)connections, particularism versus universalism and transnational solidarity. The first part elaborates on the implications of the (dis)connections introduced above by placing emphasis on rights and the media. In the second part, we discuss tensions between particularism and universalism, how they are bridged by indigenous peoples, as well as the implications this can have for understandings of climate change, justice and the media's role in covering these issues. In the third part we define solidarity in relation to justice, responsibility and agonism, detailing how solidarity is approached by interviewees and in indigenous media practices. Like responsibility, solidarity is needed for justice to be implemented.

We conclude by discussing the possibility of solidarity as a mode of communication inside and outside the media, as well as in diverse media forms. We ask: What are the implications of the (dis)connections (identified throughout this book) for actual and potential approaches to addressing the rights conflicts facing indigenous peoples—both concerning climate change and more broadly? Furthermore, what are the implications for just media practices and reporting? How can tensions detected between the particular and the universal in terms of rights, politics and media output, regarding indigenous peoples as well as climate change more broadly, be understood in relation to remedies for justice? More particularly, how can they be understood in relation to identity versus status models of recognition? Finally, how can the transnational solidarity expressed and practiced by indigenous peoples inform political processes, media practices and output? How, in other words, does transnational solidarity relate to remedies for injustice?

(Dis)connections: Rights, Politics, Media

How indigenous activists approach the politics of the COPs and the mainstream media is captured by Brown's (2002, p. 430, cf. Hanafin, 2012, p. 328) contention concerning rights: "That which we cannot *not* want is also that which ensnares us in the terms of our domination" (emphasis added). That is, indigenous groups cannot *not* want a just level of participation on a par with everyone else (i.e. recognition of status) because they are denied the right to have rights. This ensnares indigenous groups in the terms of their domination. Concurrently, they still cannot *not* want the attention of mainstream media, even though its exclusive position at the top of the attention pyramid is not as self-evident anymore. That is, they cannot *not* want the attention of the mainstream media, even if they do not trust them, because they still think, and have good reason to do so, that the mainstream media constitute connections to political elites.

For transnational groups like indigenous peoples, the disconnection between the existence of rights and the potential to access and implement them creates "a need for continuous political engagement to make real the merely declaratory nature of rights in their liberal cosmopolitan form. Such a politics is intimately related to an ethos of action, not as all-encompassing institutional praxis but as solidarity" (Hanafin, 2012, p. 329). However, if climate politics does not speak the language of the "who" (Hanafin, 2012, p. 332; see also Fraser, 2008, p. 5), claims can be made regarding the right to insubordination (Balibar, 2009, p. 303; Blanchot, 1982; Hanafin, 2012, p. 334). Such moments can occur when there is degeneration or perversion of political authority and the law. Take, for example, cases where torture is used or war is declared for purposes other than defending the nation. The role of the who or the subject (subject of law, subject of power, etc.) is however eliminated through the act of insubordination, which can take the form of a protest, and/or a withdrawal from the unfair situation (Balibar, 2009, p. 297; Hanafin, 2012, p. 334). As Blanchot (2004, pp. 41, 42; in Hanafin 2012, p. 335) explains, "When the state provokes or allows an oppressive force to threaten essential liberties, then every citizen has the right to refuse and denounce it." The way nation-states let climate change effect the lives of indigenous groups can be viewed in this light. If groups in such situations refuse or denounce what is stipulated by current political structures, a gathering of people to speak out against injustice can occur, constituting a "community that sees *justice* and *responsibility* as reasons for being together" (Hanafin, 2012, p. 335, emphasis added).

This is precisely what seems to be happening at the climate summits and other gatherings of indigenous peoples around environmental issues, sometimes in conjunction with non-indigenous people, who all have been subjected to environmental and climate injustice, and to climate change effects in general. When groups of individuals come together like this "to refuse the symbolic disembodiment which the social contract offers them, they expose the limits of law, its internal paradoxes, and the impersonality of the person with rights" (Hanafin, 2012, p. 337). The spaces created by these types of efforts are not utopian political spaces, but are constituted through "actually existing political struggle" (Hanafin, 2012, p. 338). Thus, we must again ask: who is the who of injustice? It is a person with a name, a face and a story (Cavarero, 2002, p. 99) that can reach larger publics mainly via the media.

In the introduction to this chapter, we identified several disconnections between indigenous voices and certain media forms (see also Chapters 4 and 5), particularly the mainstream or legacy media. While indigenous activists have many opportunities to speak and be visible in social media, via their organizations or on platforms like YouTube and Twitter, we argue that a crucial defining feature of mainstream media is the ability to hold politicians and other powerful actors to account. That is, the mainstream media still have the power to solicit interviews with powerful stakeholders. Thereby mainstream media could effectively hold these stakeholders responsible for justice, while at the same time shouldering their own responsibility for justice. This is a level of access that indigenous organizations and people posting on social media generally lack. Legacy media could at the same time learn from the way indigenous media organizations work with social media to connect discourses on politics, culture and spirituality, as well as from the connections that they make to a wider range of voices (both geographically and socially).

There are several methods of bringing more indigenous voices into the mainstream media: the provision of more access to national public broadcasting; more (state) support for indigenous produced media; better access to non-indigenous media; and journalism education and training programs designed to heighten public awareness of indigenous history, rights and relations, as well as human rights and justice issues in general.

While media are central to responsibility for justice, it is crucial to remember that indigenous peoples are often disconnected from influence and centrality in media and politics. However indigenous activists are making connections between different social spheres, geographical scales, and distinct groups of people. Cultural and political spheres are intertwined in many indigenous activities; for instance jointly permeating the organization of events in the Indigenous Peoples' Pavilion in the Climate Generations space at COP21 (Chapters 3 and 5). Indigenous activists combined political activism and artistic expression, using rap, film, traditional song and dance, photo exhibits, and more (Chapter 5). On social media platforms, and in co-productions with non-indigenous NGOs, indigenous produced media brought together cultural and political spheres, as illustrated for instance by an Indigenous Environmental Network video that we consider a form of self-edited media witnessing (Chapter 5). Mainstream media, on the other hand, tend to separate politics and culture, similar to how politics are disconnected from spirituality (Roosvall & Tegelberg, 2013). This has resulted in representations of indigenous peoples in de-politicized roles, appearing instead, at worst, as reified spectacles; although this seems to be changing, at least in prominent Swedish newspapers (Chapter 5). Emotion, affect and the political are largely disconnected in the mainstream media, while often strongly connected and expressed in indigenous produced (social) media (Chapter 5). Indigenous peoples are moreover connecting the particular and the universal, which we discuss further in the next section.

The Particular Versus the Universal: Rights, Politics, Media

Seila Benhabib (2007, p. 30) writes that the universalistic extension of civil and social rights, as well as of political participation rights, is constituting a new institution of citizenship, one that decouples citizenship from national belonging and from being rooted in a particular cultural community. She underlines that the rise of political activism on behalf of or by non-nationals, post-nationals, and ex-colonials shows that political agency can appear beyond the member/non-member divide. Arendt's paradoxes of the "right to have rights" are thus ameliorated by activists who "exercise their democratic-republican participation rights with or without the correct papers" (Benhabib, 2007, p. 30). Thus she emphasizes the opportunities of the universal. At the same time the universal is continually challenged and renegotiated in particular contexts, for instance by local political communities, whose self-understanding may in turn be re-negotiated in the process (Benhabib, 2006).

Indigenous activists are connecting the particular and the universal, thereby constituting a transnational solidarity that does not eschew differences. Indigenous peoples of course fall under the Universal Declaration of Human Rights (UDHR), which pertains to all peoples. The same goes for the International Covenant on Civil and Political Rights. The latter adheres largely to the logic of nation-states. Indigenous groups however form identity within *and* across nations. Identity is also strongly connected to status; if you identify as indigenous it should affect your status by granting you particular rights. This is connected to *particular* status through treaties on the national level (in some countries), and by the Convention Concerning Indigenous and Tribal Peoples in Independent Countries (UN ILO No. 169), regarding global indigenous rights.[1] In 2007, the UN General Assembly moreover adopted the United Nations Declaration on the Rights of Indigenous Peoples (UNDRIP). The significance of the Declaration is that it provides an instrument in international law that calls for the recognition of indigenous rights. The Declaration includes articles that speak directly to the importance of access to both the media and decision-making arenas in affirming the rights of indigenous peoples, for instance:

> Article 16
>
> 1. Indigenous peoples have the right to establish their own media in their own languages and to have access to all forms of non-indigenous media without discrimination.
>
> 2. States shall take effective measures to ensure that state-owned media duly reflect indigenous cultural diversity. States, without prejudice to ensuring full freedom of expression, should encourage privately owned media to adequately reflect indigenous cultural diversity.
>
> Article 18
>
> Indigenous peoples have the right to participate in decision-making in matters which would affect their rights, through representatives chosen by themselves in accordance with their own procedures, as well as to maintain and develop their own indigenous decision-making institutions.

Indigenous peoples are also specifically mentioned and protected in some paragraphs of the UNESCO Convention on the Protection and Promotion of the Diversity of Cultural Expressions. While all these rights seem to guarantee indigenous peoples just participation in (climate) politics as well as just representation in and access to the media, in reality indigenous identity often exempts people from the rights to representation (political) (Chapters 3 and 5), and to a large extent exempts them from (just) representation in the mainstream media (Chapters 4 and 5). A stronger emphasis on not only identity, but also the status of indigenous peoples, in politics and in the media, could help remedy this injustice.

In Figure 6.1 we draw on our findings and theorizations from the previous chapters, detailing how expressions of particularism and universalism relate to identity and status respectively when recognition is employed as a remedy for injustice (see also Roosvall, 2014a).

The experiences and media representations of indigenous peoples intersect with all four squares in Figure 6.1. In fact, indigenous peoples bridge and connect the squares, as illustrated in the figure. When the Identity Model meets particularism, reification of identities takes place in media reporting if the status model is not evoked at the same time. When the Status Model meets particularistic approaches the discourse becomes more political and focused on participation. Fraser (2000, p. 116) acknowledges the possibility of universalist recognition connected to the status model, which does not (like the identity model) accord "an a priori privilege to approaches that valorise group specificity." When the Identity Model in turn meets universalistic approaches a Common Human Identity (CHI) can be constructed, including indigenous and non-indigenous peoples. This identity is however de-politicized if not combined with universalistic status. Universalistic status is of course evoked when the Status model meets universalism, and global justice appears together with parity of participation worldwide. A Common Human Status (CHS) would thereby be created.

The table shows that justice is more *explicitly* connected to the status model; specifically the remedy of representation—that is just political participation. While the status model is part of a recognition approach (Fraser, 2000, 2008), it bridges the gap between this approach and the representation approach to justice. At the same time, identity aspects should not be eschewed. They should be respected according to the above-mentioned covenants, conventions and declarations. However, the reification of identities that sometimes characterizes media reporting tends to obscure political aspects. Take,

	Particularism	Universalism
Identity Model	***General approach:*** Recognition of culture and language, group identities (that could together form a larger CHI), solidarity within and between groups, the right to have rights (that is, rights regarding the recognition of identity). ***Media:*** Reification of groups in the media, de-politicization of political issues in the media; mainly in mainstream legacy media, but less in later years. ***UN conventions etc.:*** e.g. ILO169, UNDRIP.	***General approach:*** CHI, planetary solidarity. ***Media:*** De-politicization of the universality of climate change and of humanity (acknowledging universalism in vulnerability, while disregarding difference in living conditions and capabilities); mainly in mainstream legacy media. ***UN conventions etc.:*** e.g. UDHR, Convention on the Protection and Promotion of the Diversity of Cultural Expressions.
Status Model	***General approach:*** Particular rights focused on restitution, the right to insubordination, agonistic cosmopolitics, bordering on representation (as a way to guarantee recognition of status). ***Media:*** Focus on particular status of indigenous peoples, or the abuse of such status; in self-produced indigenous accounts (on social media and in national indigenous media), and in a few example articles from mainstream media (see Chapter 4 and 5). ***UN conventions etc.:*** e.g. ILO 169, UNDRIP.	***General approach:*** Global justice, parity of participation (political representation, plus redistribution and recognition), CHS - Common Human Status. ***Media:*** Focus on people from around the world coming together in political protest action; in indigenous and social media (Chapter 4 and 5); acknowledging universalism in vulnerability of humans as a presupposition but directing attention to difference in conditions, that is natural as well as political conditions (in self-produced indigenous accounts, and in a few examples from mainstream media, see Chapter 4 and 5). ***UN conventions etc.:*** e.g. UDHR, Covenant on Civil and Political Rights.

Figure 6.1. Indigenous peoples, politics and the media: particularism and universalism intersecting with identity versus status models of recognition.

for instance, news stories that focus on traditional clothing while disregarding political demands (Chapter 5). Indigenous activists called for media to focus on political aspects in our interviews, as well as in many public appearances and in their own media production. Such a focus is needed to implement justice beyond cultural rights by connecting the three main remedies for justice (Chapter 2): redistribution, recognition and representation, which must all be combined for justice to be guaranteed.

What implications does this have for media reporting? What implications does it have for broader understandings of and processes surrounding climate change, justice, the media, indigenous peoples and their potential to take part in politics? The particular and the universal have been implicitly evoked in our discussion of scales of climate justice (Chapters 2 and 3), and in our discussions concerning the rights of indigenous peoples at the COPs and beyond (Chapter 5). Here we conclude that indigenous peoples are bridging the particular and the universal, as well as the identity and status models. They embody the intersection of particular and universal rights, of particular and universal status, of particular and universal identity. Their bridging of these aspects must be taken into consideration in media reporting which would thus need to be pluralistic in terms of perspectives and voices. The bridging of these aspects could also inspire ways of approaching the universal and particular phenomenon of climate change through consideration of political as well as cultural aspects of its effects (Chapter 3) in the media and in climate politics. When relating to disconnections, the re-appropriation of speech can help us surpass territory, nation-state and identity (Hanafin, 2012, p. 332, cf. Cavarero, 2005). To be speaking can be connected to status in this regard. Hence, wording in the final document at COP21 was an important issue for several interviewees. In the subsequent examples, interviewees make statements concerning the importance of having particular formulations in the final Paris Agreement.

> First and foremost I hope that there will be *an agreement that is legally* binding, and I wish for all countries, parties to sign it. I hope there will be a decision also. I hope within the decision and agreement the *language* of the rights of indigenous peoples will be in it and not in the preambular section, since it's not legally binding. I wish also for the *word* of science and indigenous peoples' traditional knowledge to be in one paragraph. (Jannie Staffansson, Sami, Samiland, Sweden, COP21, emphasis added)

> Unfortunately, there is not so many hopes for indigenous peoples to write indigenous peoples' rights *language*, strong *language*, into this convention because there were many [past] negotiations, [with] many states against this *language*. (Dmitry Berezhkov, Itelmen people, Russia, COP21)

> We are lobbying governments to actually have … the rights of the indigenous peoples … in the outcome document, in the declaration or the agreement out of COP21. We are really lobbying to ensure that the rights of the indigenous peoples are *mentioned* in there. (Mikaele Maiava, Tokelauan, Tokelau, COP21)

Jannie Staffanssson expresses a desire to connect science and indigenous peoples' traditional knowledge, combined with a call for a general re-appropriation of language concerning the rights of indigenous peoples in the final agreement. Moreover, each of these statements emphasize not only the lack of proper language but also express a closely connected sense of a lack of rights, voice and access to political participation (Chapter 5). The same lack occurs generally in the media, where language could, but seldom does, help us surpass territory, nation-state and identity. Ultimately, these calls for legally binding language around indigenous rights were not adopted in the Paris Agreement. An agonistic cosmopolitics could help tackle this as it is "committed to the perpetual generation of new sites of action in concert on behalf of worlds not yet built or on behalf of those still emergent and in need of activist support and sustenance" (Honig, 2009, p. 133, cf. Hanafin, 2012, p. 327). While indigenous rights were not included in the agreement reached at COP21, such worlds are gradually being built by indigenous activists who bridge the particular and the universal whilst struggling for climate justice.

Solidarity and/as Communication

The existence of rights is, as mentioned, rarely enough for rights to be implemented, or even to be accessible for subjects of injustice (Arendt, 1998). Therefore responsibility is needed to "support" justice (Young, 2013, see also Chapter 2). Another supporting concept is solidarity, often associated with *collective* responsibility (Smiley, 2010). Even though media and justice have rarely been explicitly studied together (although see e.g. Boltanski, 1999; Couldry, 2012; Rao & Wasserman, 2015; Roosvall, 2016; Roosvall & Tegelberg, 2015; Silverstone, 2007), the intersection of the two is often integrated or implied in research on media and solidarity (e.g. Chouliaraki, 2011, 2013; Roosvall, 2009, 2014b).

Understandings of Solidarity

Understandings of solidarity can largely be divided into *solidarity as salvation* and *solidarity as revolution* (Chouliaraki, 2013, p. 10). Solidarity as salvation is approximated to humanitarianism (Chouliaraki, 2013, p. 11), with its emphasis on "do-gooding" (Richey, 2016). It is associated with charity, philanthropy, and other "less political" approaches (Chouliaraki, 2013, p. 11). Solidarity as revolution, conversely, is defined by its proximity to justice and political struggle. Solidarity as revolution is the strand most clearly evoked by indigenous activists calling for climate justice, and thus the one we emphasize here. Nonetheless, we view the salvation-revolution relation as a continuum (rather than a dichotomy), with features of humanitarianism possibly appearing in conjunction with revolutionary approaches.

When indigenous peoples are the subjects of climate change, which is specifically emphasized in the mainstream media material we examined (especially in 2009), they are positioned as in need of humanitarian salvation. This relates to understandings of solidarity as salvation and the world as a natural rather than a political space (Chapter 5). By contrast, when indigenous peoples are positioned as subjects of climate *injustice*, in select articles from the 2015 media material and in un-/self-edited media witnessing expressed on Facebook and YouTube, understandings of solidarity as revolution are evoked with redistribution, recognition and representation foregrounded. Jannie Staffansson calls for civil society to rise up in an act of solidarity as revolution: "We have done what we can under these negotiations. And we now call for civil society to rise. Push your ministers! We need human rights, indigenous peoples' rights in that agreement in order to stop violations, killings, murders, land grabbing, and to save the planet for our future" (Chapter 5). Staffansson underlines that the situation is urgent, which calls for normative approaches to media, democracy and politics. In other words, these are matters of our very existence that concern all of humanity, even if they are felt more concretely and urgently by some groups than they are by others.

Solidarity as revolution is in tune with agonistic approaches to democracy, highlighting ideal modes of argumentation that can allow democracy to function at its best. Chouliaraki (2013) distinguishes three forms solidarity can take in relation to the media: irony, theatricality and agonism. The form most applicable for our study is solidarity as agonism, which entails: "the communication of human vulnerability as a *political question* of *injustice* that can become the object of our *collective* reflection, *empathic* emotion and transformative *action*" (Chouliaraki, 2011, p. 377, emphasis added). Several concepts that are crucial in defining solidarity, as we understand it, are found in this quote (italicized). Solidarity is a *political* issue, connected to issues of *injustice* and calls for *collective* rather than individual responsibility. Furthermore, it builds on *empathy* and is associated with *action* rather than mere phrases (see also Liedman, 1999). Additionally, we understand solidarity to include instances of reciprocity and restitution (Kössler & Melber, 2007). Restitution connects to the *representational aspect of communicative action* (Roosvall, 2016). It refers to the recognition of historical injustice as a basis for action (including the restitution of more equal power relations). It is retroactive and re-distributional, and thereby relates to advocacy of *historical* responsibility in justice theory (Nussbaum, 2013).

Reciprocity, in turn, signifies dialogic *modes of communication* as communicative action (Roosvall, 2014b). Thus, solidarity is not a one-way form of communication, like charity (Thörn, 2008), but a two-way mode. In this view, solidarity is placed closer to revolution than it is to the salvation model described above. However, while the "other" in a dialogue is expected to be capable of contributing something that would be of interest for the "self" (again, as opposed to a one-directional charity approach), the "other" *is not required to* reciprocate (Chouliaraki, 2011, p. 263). The point is that "they" are seen as being equally interesting as potential contributors—as a "you" in agonistic cosmopolitics terms—while unequal prerequisites for action are recognized. Solidarity as a mode of communication actually stretches beyond dialogue in that exchange should be similar or equal (not true of all dialogue), speaking to a "you" or "another" rather than the "Other", and in that it includes actions directed towards injustice (Roosvall, 2014b, 2016). Thereby, it surpasses ideas on dialogue leading to peace/agreement (as in deliberative approaches, see Chapter 5) since everyone will hardly agree upon the mending of injustice. The plurality of indigenous activist voices that question power relations, access to political participation and dominant ways of responding to climate change and justice constitutes one such approach.

Transnational Solidarity in Indigenous Activism and Journalism: Attitudes and Practices

Several interviewees use the solidarity concept to describe their activism and how it connects indigenous activists from around the world. Solidarity has traditionally been closely connected to the nation (Renan, 1882/1990), but today solidarity with distant others is the focus of theorization and investigation in relation to the media (Boltanski, 1999; Calhoun, 2002, Chouliaraki, 2013; Orgad, 2012). Indigenous activists combine this solidarity towards their own nation with solidarity towards other indigenous peoples, as well as towards humanity at large. Thereby, they are not relating to distant others but rather to a "you" or "another", which is the signifier of an agonistic cosmopolitics of human rights (Cavarero, 2000; Hanafin, 2012).

As Anthony Debbarma puts it:

> [T]he Asia Indigenous Peoples Pact is one platform. We also collaborate with African indigenous people, with Latin America, yeah with every … regions of the world, we come together and of course this is one of the pavilions where we come together and we share together our *solidarity* and our concerns.[2] (Anthony Debbarma, Borok people, India, COP21, emphasis added)

In the quote introducing this chapter, India Logan-Riley relates to co-operation in a similar way, without explicitly mentioning the word solidarity. Logan-Riley underlines collectivity and thereby connects to a crucial aspect of solidarity: collective responsibility (Smiley, 2010). She relates this furthermore to colonial experiences, thereby politicizing the practice of solidarity and thus evoking solidarity as revolution (rather than more humanitarian, non-political approaches). Logan-Riley also mentions struggle, which expresses an agonistic approach. Like Debbarma, she talks about indigenous peoples in general, not just her own indigenous group.

Lene Kielsen Holm brings up solidarity to describe how the Indigenous Caucus works at the summits:

> [T]he globe is divided into seven indigenous regions … for instance, for the Arctic, it is the Inuit and the Sami that is in the same caucus. While Asia has its own, the Pacific, its own … South America or Latin America and the Caribbean, North America … So we are divided into those different ones. But, of course, in *solidarity* with each other because we are all indigenous peoples and … to a great extent, we are in the same situation. (Lene Kielsen Holm, Inuit, Greenland, COP21, emphasis added)

When an indigenous group lacked access to certain negotiations, other indigenous representatives with access would share information (Chapters 3 and 5). Holm associates this practice with solidarity, a solidarity that is transnational (like many of the indigenous groups). Allison Akootchook Warden spoke in a similar manner of a solidarity through connections with other indigenous and non-indigenous peoples:

> I hope to make new friends with activists and coalition people from all over the world, and build relationships with people here in Paris and other people who have come from all over the world that are long-lasting and meaningful and *reciprocal*. And I hope to support what they are working on in their hometown … all over the planet

> people are facing different issues and to protect their earth, and I'm hoping to stand in *solidarity* and learn about what they are fighting for and hope they will stand in *solidarity* to protect my home and my people. (Allison Akootchook Warden, Iñupiaq people, USA, COP21, emphasis added)

Warden echoes an understanding of solidarity as reciprocal in her view on relationships between indigenous groups. Her statement concurrently evokes a bridging between the *particular* and the *universal*. She stresses difference as well as commonality, talks about making *connections* with other activists, and alludes to existing connections with indigenous peoples around the world. When asked what role climate justice activists should play inside or outside the summit, Warden extended solidarity to the whole planet:

> I think raising awareness of the impacts of climate change that are happening in their communities now, connecting it to actual fights that are happening on the ground to prevent drilling of resources and pipelines and so forth all over the world. We need to keep 80 percent of the oil and the things that create carbon in the ground now in order to, you know, live a life on this planet, and our role is to create awareness and to help strengthen the movement so that more people … who might not have ever heard of this will feel inspired to take action and to stand in *solidarity* with the people on the *planet*, and to fight for our *planet*. (Allison Akootchook Warden, Iñupiaq people, USA, COP21, emphasis added)

The planetary solidarity evoked by Warden corresponds—together with her statements on solidarity for certain communities—with Kurasawa's (2007, p. 177, cf. Hanafin, 2012, p. 329) view on cosmopolitan solidarity as "a networked practice of a vast web of actors who labour to create nodes of commonality and points of intersection out of shifting, cross-cutting lines of affinity that remain grounded in local and national settings."

Transnational solidarity is also performed by indigenous media. For example, Pirita Näkkäläjärvi, Head of Yle Sápmi, the indigenous division of Finnish national broadcasting, describes the sharing of material between Sami media across Finland, Sweden and Norway.

> Unfortunately we did not have resources to send a reporter to Paris, so we had to work from Inari [in Finland]. During COP21, we cooperated with Sameradion and SVT Sapmi [indigenous divisions of national broadcasting in Sweden] and NRK Sapmi [indigenous division of national broadcasting in Norway], which were able to send reporters to Paris. This is often the … case: they are able to go … for example to the UN's Permanent Forum on Indigenous Issues in New York City but we stay home and take advantage of their materials especially for our radio programmes and website.

> NRK Sapmi is very well resourced (staff of 105 compared to our 20) and Swedish colleagues are able to apply for travel grants, which we are not able to do as part of tax-funded Finnish Broadcasting Company. The quality of our reporting on indigenous issues, indigenous events and for example COP21 was and is seriously deteriorated by the fact that we are not able to do first hand reporting from these major international events. (Pirita Näkkäläjärvi, Sami, Samiland, Finland, Head of Yle Sápmi, e-mail interview 2017)

Unequal conditions of production for Sami media in Nordic countries seems to motivate the solidarian action of sharing. Aside from this sharing of material, Näkkäläjärvi explained that stories produced by NRK Sapmi, Sameradion and SVT Sapmi were aired in the joint Nordic Sami TV news (Ođđasat), and as part of their own programming for Yle Sami Radio in Finland. Yle Sápmi is also an associate partner of the World Indigenous Television Broadcasters Network. However, because NRK Sapmi has more resources, it is they who handle the cooperation.

Curtis Konek, of the Canadian Inuit Youth Delegation, suggested that sharing media material about climate change should become a common practice since the issue is acute (Curtis Konek, Inuit, Canada, COP17). He emphasized that particularly media organizations with the ability to reach large populations should engage in sharing material. This is, of course, not likely to happen on a large scale in a competitive media market. However, it has not stopped indigenous media producers from doing it anyway. The latter, together with the collective actions of indigenous activists, constitutes an agonistic cosmopolitics; that is a "plural and interactive space of exhibition that is the only space that deserves the name of politics" (Cavarero, 2000, p. 57, cf. Hanafin, 2012, p. 331). For climate politics to deserve the name of politics in this sense, it should constitute such a plural and interactive space of exhibition; it should be signified by solidarian practices and attitudes.

Conclusion

Solidarity as a mode of communication expresses a solidarian *attitude* towards the subjects discussed. Such an attitude is signified by being democratic since, unlike dialogue in general, it necessarily includes the components of empathy, collective responsibility, restitution, reciprocity, action and political worldviews, at least in its revolutionary form. Solidarian modes of communication could thereby facilitate democratic *participation*, which is necessary for the implementation of justice for indigenous groups. This corresponds to the representation component of Fraser's (2008) redistribution-recognition-representation triad, and can in turn become a vehicle for assessing and obtaining a just implementation of redistribution and recognition. A solidarian mode of communication would thus enable democracy through exchange in not merely deliberative but agonistic forms, focusing on difference, restitution and ways of dealing with injustice that surpass an unconditional celebration of peace and consensus. This would simultaneously serve to facilitate participation that is connected to an inclusive politics of voice and representation (see Fenton, 2008). Thereby, issues of injustice could be properly addressed. Such measures are urgently needed when reporting on or engaging in any sort of public discourse on climate change. Journalism in nation-states that have already experienced acute climate change effects, like Bangladesh, have begun to set an example by moving towards advocacy journalism. These approaches acknowledge that the climate crisis calls for well-aware, engaged actions and language that aims at curbing climate change (see Rahman, 2014).

Solidarian modes of communication constitute such well-aware, engaged practices. Moreover, they go beyond ideas of solidarity as salvation and more basic humanitarian approaches by including the idea that others can contribute to an exchange on similar or equal terms, which mere charity or general dialogue models do not necessarily support. They also include responsibility for justice not only in our time but also for the past (and the future), through restitution and connected remedies for injustice (Nussbaum, 2013; Young, 2013). Furthermore the "mode" in the solidarian mode of communication is crucial. It implies a way of speaking *in and to* the world signified by collective responsibility, action, empathy, reciprocity and political views. By speaking in and to the world the other is addressed as another or as a you (see Cavarero, 2009; Hanafin, 2012). The development of solidarian modes of communication can therefore contribute to enhancing the methods and conditions of agonistic debate, discussion and persuasion. This is an essential measure for journalism and society at large, especially regarding the acute issue of climate change and more specifically climate justice.

Solidarian modes of communication resemble "global journalism" as a mode of communication (Berglez, 2013). Beyond technical means, global journalism is constituted by an epistemological component: it is based on a global mode of thinking about society (cognition) and language use (discourse) (Berglez, 2013, p. 97). Solidarian modes of communication also exist beyond the roles of technology. In other words, instead of being constituted by technology, solidarity as a mode of communication can, like global journalism, be understood as epistemologically constituted through a solidarian mode of thinking about society (cognition) and through connected solidarian language use (discourse). It is thus constituted by collective rather than connective action (but can work together with connective action, as shown in Chapter 4) (Bennett & Segerberg, 2013; Segerberg & Bennett, 2011). A solidarian mode of communication responds to how global conditions and complex relations can be represented in the news. It also encompasses a plurality of voices, organizing them in an interactive way for news communication (Berglez, 2013, p. 110). It does so through scalar transcendence (Christensen, 2013), which at times extends to the global level, but does not necessarily require this (as global journalism would). Non-global transnational transcendence, in other words, is enough. Thus, in sum, a solidarian mode of communication responds through an attitude (discursive mode of address) and practice (discursive or non-discursive mode of action) of reciprocal exchange, empathy, responsibility, restitution, action and politics, in a way that transcends scales; most prominently expressed by the connections indigenous activists make between the local and the transnational.

We suggest that the application of a solidarian mode of communication, in media and in politics, could help address climate change as an issue of climate justice and support the just inclusion of indigenous peoples in these discussions. If applied, it would respond not only to calls for climate justice expressed by indigenous activists at the COPs, but also to their collective calls for indigenous rights more generally. Finally, a solidarian mode of communication could help save the planet by serving to integrate the perspectives and collective practices of indigenous activists into public discourses on climate change and climate justice, framing them as essentially transnational issues with concurrently particularistic and universal implications. Suggesting a solidarian mode of communication, as it has been defined here, is on a more general level our response to what Couldry (2012, p. 210) terms the underlying point of media studies; that is "to study how media contribute to the conditions of knowledge and agency *in the world* and so to understand better whether media enable us to live well together."

Returning to the questions raised at the beginning of this chapter, we conclude that the (dis)connections identified throughout this book for actual and potential approaches to indigenous rights conflicts, both in terms of climate change and more broadly, call for an agonistic cosmopolitics. This entails a plurality of perspectives and voices that are necessary for democracy to work in conjunction with justice. The same pluralism is required for just media reporting and practices to take place. We also note that some activists have taken an insubordination approach towards the mainstream media. The mainstream media however continues to influence political processes due to their privileged access to political elites. Thereby the mainstream media still needs to modify their practices and learn from indigenous peoples and from social media when it comes to the inclusion of voices, perspectives, and knowledge about indigenous peoples, rights and justice.

The tensions between the particular and the universal, found in media reporting as well as in approaches towards indigenous peoples in politics, imply a conflict between the national and the transnational that permeates much of contemporary public discourse (beyond climate politics). Understanding these tensions entails relating them to identity versus status. Problems that come with universal identity—potential de-politicization—can be avoided when we also focus on universal status, which is inherently political and implies the need for political participation, or representation in Fraser's (2008) terms. Problems associated with the media's focus on particular identities, such as the reification of people and disregard for the political, can be avoided by employing particularistic discourses in relation to status, guided for instance by the UNDRIP. At the same time identity should not be eschewed. Recognition of identity has provided crucial insights regarding racism (Fraser, 2000, p. 113), which is relevant for indigenous groups in relation to politics and to the media.

Disconnections between the particular and the universal are bridged by indigenous peoples in terms of their concurrently particular and universal rights, in terms of their scalar transcendence in activism, and in many cases as people (transnational groups). They also embody a connection between identity and status since indigenous identity is the basis for indigenous status; identifying as indigenous ideally brings status (if grounded in the characteristics of indigenous peoples, see Chapter 1). The bridging work that is done by indigenous activists, indigenous media and indigenous peoples more generally is a model for climate politics in that it employs scalar transcendence, and includes all geographical scales of climate justice: the intranational, the international, and the transnational. Such a scope is necessary for global climate justice to be employed (as we concluded in Chapter 2). Global climate justice cannot build on justice frameworks exclusively established between nation-states. Transnational and intranational scales must be included too. This transcendence also vouches for the inclusion of multiple perspectives and voices. Thus *recognition*, applied in this way, facilitates representation.

At the beginning of this chapter we also asked: how can the transnational solidarity expressed and performed by indigenous peoples more specifically inform political processes as well as media output and practice, and how does it relate to remedies for injustice? Solidarian modes of communication could, as mentioned above, facilitate democratic *participation*, which is necessary for the implementation of justice for indigenous groups. The way solidarity is understood and performed by the interviewed indigenous activists, and by indigenous media, implies that solidarity must concurrently be intra- and transnational in order to correspond to phenomena and groups that are likewise intra- and transnational: namely, climate change and indigenous peoples. A solidarian mode of communication would enable democracy through multiscalar (Widholm, 2011) exchange. It would do this in agonistic forms, surpassing the deliberative goal of peace and consensus, by instead focusing on restitution (of previous wrongs) and on differences regarding conditions (see Chapter 3 on natural conditions, Chapter 4 on media-related conditions, and Chapter 5 on political conditions). Participation that is connected to an inclusive politics of voice and representation (see Fenton, 2008) in a multiscalar way is thereby facilitated. Thus, issues of injustice are addressed and the *representation* component of Fraser's (2008) Redistribution-Recognition-Representation triad works as a vehicle for also tackling redistribution and recognition. This book has revealed that taking such measures are urgently needed for indigenous peoples.

Finally, returning to the main question posed in Chapter 1, in this book we have demonstrated the roles different types of media play simultaneously as agents and arenas in relation to climate justice and transnational climate justice activism. We have also shown how they can learn from one another's practices. For instance, mainstream media could learn from intersections with social and indigenous media how to provide adequate knowledge on indigenous peoples (identity, culture, natural and political conditions, status). They could also find ways to include multiple voices and perspectives (as in agonistic cosmopolitics) using multiscalar and transnational approaches, thus relating to transnational climate justice and to justice for indigenous peoples more broadly. One role remains that pertains only to the mainstream media, and that is difficult for social media or state-supported indigenous media to embody: that of conversation partners with political elites that are capable of holding them to account. When this is combined with the fact that mainstream media remain essential information sources for political elites, and continue to reach large national publics that are crucial to political communication within the nation-state, it becomes evident that mainstream media continue to be fundamental in changing media ecologies. Redistribution of means is however needed in the media sphere as well as in politics in order for justice to be applied. More support for indigenous sources, whether on social media platforms or not, would also help implement recognition and representation. This would ultimately (as concluded in Chapter 2) be facilitated by a focus on intra- and transnational (climate) justice, surpassing the largely one-sided international emphasis that has characterized climate justice discourses to date. Heeding intra- and transnational calls for climate justice, expressed by indigenous activists and indigenous peoples more broadly, thus contributes to a better developed understanding of justice in a globalizing age.

Notes

1. The Convention has been ratified by 22 countries since it was established in 1989. The international convention is an important precursor to UNDRIP.
2. Debbarma refers to the Indigenous Peoples' Pavilion in the Generations Space at COP21 in Paris.

References

Arendt, H. (1998). *The human condition* (2nd ed.). Chicago, IL: University of Chicago Press.

Balibar, E. (2009). Blanchot l'insoumis. In M. Antselme, G. Berkman, C. Bident, J. Degeneve, L. Hill, M. Holland, … P. Shahrjerd (Eds.), *Blanchot dans son siècle*. Paris: Sens Public.

Benhabib, S. (2006). *Another cosmopolitanism*. London: Oxford University Press.

Benhabib, S. (2007). Another universalism: On the unity and diversity of human rights. *Proceedings and Addresses of the American Philosophical Association, 81*(2), 7–32.

Bennett, W. L., & Segerberg, A. (2013). *The logic of connective action: Digital media and the personalization of contentious politics*. New York, NY: Cambridge University Press.

Berglez, P. (2013). *Global journalism: Theory and practice*. New York, NY: Peter Lang.

Blanchot, M. (1982). *The space of literature*. Lincoln, NE: University of Nebraska Press.

Blanchot, M. (2004). Preciazione. In M. Blanchot (Ed.), Nostra compagna clandestine: Scritti politici (1958–1993) (pp. 39–42). Napoli: Cronopio.

Boltanski, L. (1999). *Distant suffering: Morality, media and politics*. Cambridge: Cambridge University Press.

Brown, W. (2002). Suffering the paradoxes of rights. In W. Brown, & J. Halley (Eds.), *Left legalism/Left critique* (pp. 420–434). Durham, NC: Duke University Press.

Calhoun, C. (2002). Imagining solidarity: Cosmopolitanism, constitutional patriotism, and the public sphere. *Public Culture, 14*(1), 147–171.

Cavarero, A. (2000). *Relating narratives: Storytelling and selfhood*. London: Routledge.

Cavarero, A. (2002). Who engenders politics? In G. Parati, & R. West (Eds.), *Italian feminist theory and practice: Equality and sexual difference* (pp. 88–103). Madison, WI: Fairleigh Dickinson University Press.

Cavarero, A. (2005). *For more than voice: Toward a philosophy of vocal expression*. Stanford, CA: Stanford University Press.

Chouliaraki, L. (2011). Improper distance: Towards a critical account of solidarity as irony. *International Journal of Cultural Studies*, *14*(4), 363–381.

Chouliaraki, L. (2013). *The ironic spectator: Solidarity in the age of post-humanitarianism*. Cambridge, MA: Polity Press.

Christensen, M. (2013). Arctic climate change and the media: The news story that was. In M. Christensen, A. E. Nilsson, & N. Wormbs (Eds.), *Media and the politics of Arctic climate change: When the ice breaks* (pp. 26–51). Basingstoke: Palgrave.

Cottle, S. (2009). Global crises in the news: Staging new wars, disasters, and climate change. *International Journal of Communication*, *3*, 494–516.

Couldry, N. (2012). *Media, society, world: Social theory and digital media practice*. Cambridge: Polity.

Fenton, N. (2008). Mediating solidarity. *Global Media and Communication*, *4*(1), 37–57.

Fraser, N. (2000). Rethinking recognition: Overcoming displacement and reification in cultural politics. *New Left Review*, *3*, 107–120.

Fraser, N. (2008). *Scales of justice: Reimagining political space in a globalizing world*. New York, NY: Columbia University Press.

Hanafin, P. (2012). A right to politics? Towards an agonistic cosmopolitics of human rights. In G. Delanty (Ed.), *Routledge handbook of cosmopolitanism studies* (pp. 326–339). Abingdon: Routledge.

Honig, B. (2009). *Emergency politics: Paradox, law, democracy*. Princeton, NJ: Princeton University Press.

Kössler, R., & Melber, H. (2007). International civil society and the challenge for global solidarity. *Development Dialogue*, *49*, 29–39.

Kurasawa, F. (2007). *The work of global justice: Human rights as practices*. Cambridge: Cambridge University Press.

Liedman, S. E. (1999). *Att se sig själv i andra. Om solidaritet*. [*To see yourself in others. On solidarity*]. Stockholm: Bonniers.

Nussbaum, M. C. (2013). Foreword. In I. M. Young (Ed.), *Responsibility for justice* (pp. ix–xxv). Oxford: Oxford University Press.

Orgad, S. (2012) *Media representation and the global imagination*. Cambridge: Polity.

Rahman, M. (2014). Justice delayed-justice denied: Framing climate justice in Bangladeshi newspaper journalism during COPs. In B. León (Ed.), *Communicating climate change: From global agenda to media representation* (pp. 179–212). Salamanca: Comunicación Social.

Rao, S., & Wasserman, H. (Eds.). (2015). *Media ethics and justice in the age of globalization*. London: Palgrave Macmillan.

Renan, E. (1882/1990). What is a Nation? In H. K. Bhabha (Ed.), *Nation and narration* (pp. 8–22). London: Routledge.

Richey, L. A. (2016). Introduction: Celebrity humanitarianism and north-south relations—Politics, place and power. In L. A. Richey (Ed.), *Celebrity humanitarianism and north south relations: Politics, place and power* (pp. 1–24). Abingdon: Routledge.

Roosvall, A. (2009). Global divides in cosmographic genres: Charity, solidarity and different explanations of difference. *Nordicom Review*, *30*(1), 163–174.

Roosvall, A. (2014a). The identity politics of world news: Oneness, particularity, identity and status in online slideshows. *International Journal of Cultural Studies*, 17(1), pp. 55–74.

Roosvall, A. (2014b). Beyond dialogue: Exploring solidarity as a mode of communication through a debate on readers' comments to online news. *Northern Lights*, *12*(1), 49–67.

Roosvall, A. (2016). Journalism, climate change, justice and solidarity: Exploring attitudes in editorials surrounding the IPCC reports. In E. Eide, R. Kunelius, M. Tegelberg, & D. Yagodin (Eds.), *Media and global climate knowledge: Climate journalism and the IPCC* (pp. 129–150). Basingstoke: Palgrave.

Roosvall, A., & Tegelberg, M. (2013). Framing climate change and indigenous peoples: Intermediaries of urgency, spirituality and de-nationalization. *International Communication Gazette*, *75*(4), 392–409.

Roosvall, A., & Tegelberg, M. (2015). Media and the geographies of climate justice: Indigenous peoples, nature and the geopolitics of climate change. *TripleC: Communication, Capitalism & Critique*, *13*(1), 39–54.

Segerberg, A., & Bennett, L. (2011). Social media and the organization of collective action: Using Twitter to explore the ecologies of two climate change protests. *Communication Review*, *14*, 197–215.

Silverstone, R. (2007). *Media and morality: On the rise of the mediapolis*. Cambridge: Polity.

Smiley, M. (2010). Collective responsibility. *Stanford Encyclopedia of Philosophy*. Retrieved from http://www.science.uva.nl/~seop/entries/collective-responsibility/

Thörn, H. (2008). Solidariteten betydelse(r): The meaning(s) of solidarity. *Ord & Bild*, *2*, 150–158.

Widholm, A. (2011). *Europe in transition: transnational television news and European identity*. Stockholm: JMK.

Young, I. M. (2013). *Responsibility for justice*. Oxford: Oxford University Press.

APPENDIX

Interview Questions

COP17 Interviews

Conducted on December 5 and December 6, 2011. Durban, South Africa.

Questions

1. What do you think about media coverage of indigenous peoples and climate change?
2. Is there a difference between mainstream media and alternative media reporting on indigenous peoples and climate change?
3. Do you have an opinion on how media coverage could be improved?
4. Do you have advice for media researchers on how to approach these issues in future research?

COP21 Interviews

Conducted between November 28 and December 4, 2015. Paris, France.

Questions

1. Background, Local Geographies

What is your name? Do we have consent to publish what you say in this interview, and do we have consent to use your name?

Do you represent an organization? Which one?

What is your role in that organization?

Do you belong to an indigenous community/group? Which one?

What country, and what area of that country, do you live in?

Can you tell us more about your people or your community and its experiences of climate change? What types of impacts (ecological, cultural, economic) is climate change having in your community?

2. Climate Summits and Activism/the Political

Is this your first time at a climate summit? *If not, which ones have you attended in the past and what feels different about COP21?

Do you have access to the summit halls?

Can you describe your activities here? Where do they take place, what are they, how long will you be active here?

Why are you engaged in the climate change issue?

Are you working transnationally?

How is activism vs. regular politics important for progress regarding protection of the climate and/or climate justice?

What do you hope to accomplish at COP21? What do you expect will be the summit outcomes?

3. Climate Justice

Do you use the term climate justice?

How do you define climate justice?

What does climate justice mean to you?

What role will climate justice activists play at/outside the summit?

Does media play a role in your efforts to advocate for climate justice?

4. Role of Media

How much do you follow media? What media do you read, watch, etc.?

How much do you use social media/new media?

What do you think about the media coverage of climate change?

Do you perceive a difference between mainstream and alternative media?

How could media coverage, mainstream and/or alternative, cover climate change better?

What do you think about social media's role in covering climate change and the summits?

What do you think about the role of indigenous and/or local media in covering climate change and the summits?

How are you engaged in media activities? (Seeking media attention for a cause? Using social media? Working for indigenous media? Etc.)

Interview with Sami Journalist

Pirita Näkkäläjärvi, Sami people, Samiland, Finland. Head of Yle Sápmi (indigenous division of Finnish broadcasting), Inari, Finland.
E-mail interview conducted on January 3, 2017.

Questions

1. When reporting from COP21 about issues regarding climate change for indigenous peoples, to what extent did you focus on groups from the areas where you are based and to what extent did you cover issues concerning indigenous groups living in other parts of the world? Can you give some examples?

2. Did you use the term climate justice in your reporting? If so, do you remember/can you give examples of what context(s) you used it in?

3. Did you use the term justice more broadly in your reporting? If so, do you remember/can you give examples on what context(s) you used it in?

4. How do you think your reporting from COP21 differed from reporting done by the mainstream media?

5. Did you focus on both the traditional politics and negotiations of the COP, on one hand, and on activism carried out by indigenous peoples on the other? If not, what did you focus on and why? If you did focus on both, to what extent and why did you do that?

6. What was your most important story from COP21? Why?

7. Did mainstream media pick up on some of your stories? Which ones?

8. Do you strive to have your stories picked up by mainstream media? Why (not)? If you do, what strategies do you use?

9. Do you work as both an activist and a journalist? If so is that problematic in any way? Why is that (not) so?

10. What media do you trust, like and/or rely on when it comes to covering climate change and indigenous peoples at the COPs?

11. Do you co-operate with other indigenous based media organizations? In that case which ones?

12. What is the main outlet for your coverage (YouTube, your own website, Facebook, etc.)?

13. How much do you use social media in your work?

14. Do you intend to change the way you work at COP22 in Marrakesh in some way based on your experiences from COP21 in Paris?

INDEX

A

B

C

E

F

G

H

J

K

L

M

N

O

P

Q

R

U

V

W

Y

Z

GLOBAL CRISES AND THE MEDIA

Simon Cottle, *General Editor*

From climate change to the war on terror, financial meltdowns to forced migrations, pandemics to world poverty, and humanitarian disasters to the denial of human rights, these and other crises represent the dark side of our globalized planet. They are endemic to the contemporary global world and so too are they highly dependent on the world's media.

Each of the specially commissioned books in the *Global Crises and the Media* series examines the media's role, representation, and responsibility in covering major global crises. They show how the media can enter into their constitution, enacting them on the public stage and thereby helping to shape their future trajectory around the world. Each book provides a sophisticated and empirically engaged understanding of the topic in order to invigorate the wider academic study and public debate about the most pressing and historically unprecedented global crises of our time.

For further information about the series and submitting manuscripts, please contact:

Dr. Simon Cottle
Cardiff School of Journalism
Cardiff University, Room 1.28
The Bute Building, King Edward VII Ave.
Cardiff CF10 3NB
United Kingdom
CottleS@cardiff.ac.uk

www.ingramcontent.com/pod-product-compliance
Ingram Content Group UK Ltd.
Pitfield, Milton Keynes, MK11 3LW, UK
UKHW020719041026
15590UKWH00002B/761

9 781433 134876